RESEARCH METHODS FOR PRODUCT DESIGN

ALEX MILTON & PAUL RODGERS

SECOND EDITION

Laurence King

First published in 2013
This edition published in Great Britain in 2026 by

LAURENCE KING

Laurence King
An imprint of Quercus
Carmelite House
50 Victoria Embankment
London EC4Y 0DZ

An Hachette UK company
The authorised representative in the EEA is Hachette Ireland,
8 Castlecourt Centre, Dublin 15, D15 XTP3, Ireland
(email: info@hbgi.ie)

Copyright © text 2026 Alex Milton and Paul Rodgers

The moral right of Alex Milton and Paul Rodgers to be identified as the authors of this work has been asserted in accordance with the Copyright, Designs and Patents Act, 1988.

All rights reserved. No part of this publication may be reproduced or transmitted in any form or by any means, electronic or mechanical, including photocopy, recording, or any information storage and retrieval system, without permission in writing from the publisher.

A CIP catalogue record for this book is available from the British Library

TPB ISBN 978-1-52943-600-6
Ebook ISBN 978-1-52943-601-3

Quercus hereby exclude all liability to the extent permitted by law for any errors or omissions in this book and for any loss, damage or expense (whether direct or indirect) suffered by a third party relying on any information contained in this book.

10 9 8 7 6 5 4 3 2 1

Commissioning editor: Kara Hattersley-Smith
Designer for this edition: hoopdesign.co.uk
Project manager: Catherine Hooper
Picture researcher: Maria Ranauro

Printed and bound in China by C&C Offset Printing Co., Ltd.

Papers used by Quercus are from well-managed forests and other responsible sources.

CONTENTS

6 INTRODUCTION

1 WHAT IS PRODUCT DESIGN RESEARCH?

- **9** What is research?
- **9** What is design research?
- **10** Primary and secondary research
- **11** The iterative design research process
- **12** Product design process and methods
- **14** Summary of research methods involved in the design process
- **16** Analysing research
- **17** Ethics
- **19** Reflecting upon your research journey

2 LOOKING

- **21** Ethnography
- **24** Photo and video diaries
- **26** Shadowing
- **27** A day in the life
- **29** Personal belongings
- **30** Future forecasting
- **31** Trend spotting
- **32** Scenarios, speculations and preferable futures
- **34** Product autopsy
- **37** Sketching
- **40** CASE STUDY Jinil Park's Drawing Series
- **42** CASE STUDY Paula Zuccotti's Future Archaeology of a Global Lockdown
- **44** TUTORIAL How to conduct an ethnographic study
- **46** TUTORIAL How to conduct a day-in-the-life study

3 LEARNING

- **49** Cultural probes
- **51** Competitor product analysis
- **52** Literature reviews
- **53** Cultural observation
- **53** Internet searches and using generative AI as a research tool
- **55** AI language models as research assistant
- **56** Cultural comparisons
- **56** Role playing

58 Try it yourself
60 Mind mapping
61 Sampling
62 Task analysis
64 **CASE STUDY** delaO Design Studio's Caravana: Designing a mobile coffee unit for social impact
66 **CASE STUDY** Mischer'traxler studio's *access*: Global water distribution through glassware
68 **TUTORIAL** How to write a literature review
70 **TUTORIAL** How to create a great mind map

4 ASKING

73 Questionnaire and Surveys
74 Focus and unfocus groups
75 User narration
76 Interviews
77 Be your customer/client
78 Brand DNA analysis
80 Market and retail research
82 Image and mood boards
84 Perceptual mapping
85 Personas
86 Product collage
87 Extreme users
88 Journey mapping
88 Using and evaluating data
90 **CASE STUDY** Kate Strudwick's for.form: Redefining forensic evidence handling
92 **CASE STUDY** Parsons & Charlesworth's Catalog for the Post-Human
94 **TUTORIAL** How to write a great questionnaire
96 **TUTORIAL** How to conduct great interviews

5 MAKING

99 Sketch modelling
100 Mock-ups
102 Paper prototyping
104 Quick-and-dirty prototypes
105 Experience prototyping
106 Appearance models
108 Empathy tools
110 Bodystorming
110 Rapid prototyping
114 **CASE STUDY** 4c Design's Numnuts
116 **CASE STUDY** Pili Wu's plastic ceramics
118 **TUTORIAL** How to conduct experience prototyping
120 **TUTORIAL** How to do quick-and-dirty prototyping

6 TESTING

- 123 Scenario testing
- 124 User trials
- 126 Product usability
- 129 Material testing
- 130 Safety testing
- 132 Circular design testing
- 134 **CASE STUDY** Berghaus Freeflow: Redefining backpack comfort
- 136 **CASE STUDY** Enhancing PAPR device usability & design through task analysis
- 138 **TUTORIAL** How to run a great user trial

7 EVALUATING & SELECTING

- 141 Choosing the right methods
- 142 Checklists
- 143 External decision making
- 144 Intuition
- 146 Crowdsourcing
- 147 Product champions
- 148 Matrix evaluation
- 150 **CASE STUDY** Lotus Theory 1
- 152 **CASE STUDY** PA Consulting & Cumulus's brain health monitoring platform
- 154 **TUTORIAL** How to conduct a matrix evaluation
- 156 **TUTORIAL** How to write a checklist (PDS)

8 COMMUNICATING

- 159 Preparing a presentation
- 160 Ten tips for a great presentation
- 162 Report creation
- 164 Presentation visuals and models
- 164 Presentation guidelines
- 166 Engaging the public
- 168 **CASE STUDY** AMO/Rem Koolhaas Roadmap 2050 project
- 170 **CASE STUDY** Giovanni Innella & Gionata Gatto's GeoMerce
- 172 **TUTORIAL** How to create a great research presentation
- 174 **TUTORIAL** How to create a great research report

- 176 Summary
- 178 Glossary
- 182 Resources
- 186 Index
- 190 Picture credits
- 191 Case study credits
- 192 Acknowledgements

INTRODUCTION

Product designers need a comprehensive understanding of research methods as their day-to-day work routinely involves them observing people, asking questions, searching for information, making and testing ideas, and ultimately generating solutions to problems. The act of research is manifest in the design process. Product designers and companies fully acknowledge the importance of research in their work – indeed, the design research methods and particular approaches that a designer or company chooses will often differentiate them from others and provide distinctive advantages for their practice, organization, stakeholders and clients.

Product design is now a global phenomenon; competition for work today transcends physical, national and cultural borders, and the complex global challenges such as ageing populations, disruptive technologies, economic instabilities and inequalities, conflict and displacement, and environmental degradation and injustices mean that product designers have to offer far more in terms of research expertise than ever before.

The traditional Western perspective on design is that it shapes products to make them easier to use and manufacture, to make them more attractive to consumers, and to add status and create value. Since the first edition of this book was published in 2013 there has been a shift in how we understand and discuss design and its ability to contribute to the world we live in. Central to this transformation is how we define design and the development of new fields or contexts for design practice and research. Service design, design for social innovation and planet-centred design, along with an ever-expanding range of interdisciplinary collaborations, are all significant drivers of this paradigm shift.

Huge technological advances in information, computing and manufacturing processes also offer enormous opportunities to product designers, such as the development of 'intelligent' and integrated products, services and systems, although these also create fresh challenges and raise important research questions that need to be dealt with. Product designers are, in many ways, well placed to address these challenges because of the manner in which they apply their design thinking to problems.

This book helps you to conduct effective, ethical and useful research in order to create better products that users will find pleasure in. It also invites you to consider that design is not just an activity motivated by commercial gain: it can also be about addressing questions of reflection and change, as well as criticizing or commenting on prevailing social structures and emergent technologies.

This book demonstrates, in a clear, highly visual and structured fashion, how research methods can support product designers and help them address the very real issues the world faces today. Research methods are a somewhat neglected subject in many product design courses around the world. The key goal of this book, therefore, is to introduce students and early career practitioners to the variety of research methods and tools that can be used, as well as ideas about how and when to deploy them effectively. It contains a number of new and existing methods that will enable you to investigate people, form, lifestyles, services, tools and processes in ways that will make your work more useful and more delightful for the intended audience. The book seeks to instil the value of research, helping you to develop the skills to research, ideate, communicate and deliver products that are just and equitable, and socio-culturally, politically and environmentally aware.

Each method is illustrated in the book with real-world examples in the form of case studies, tutorials and professional advice, all backed by rich visual imagery. The book covers qualitative and quantitative research methods, ethnography, mapping, trend forecasting, cultural diversity, video diaries, cultural probes and many other methods from the wide spectrum of contemporary product design. Interspersed throughout the book are several double-page 'how to' features. These cover the key activities that form the chapter headings of the book, namely Looking, Learning, Asking, Making (Prototyping), Testing, Evaluating and Selecting, and Communicating. The chapter headings themselves provide a clear, structured and informative resource for any product designer involved in research, be that within an educational, industrial or social context.

WHY READ THIS BOOK?
This book is aimed at design students, early career practitioners and those looking for an introduction to the techniques and methods of product design research. It will provide you with a comprehensive, relevant and visually rich insight into the world of research methods specifically aimed at product designers, while also proving useful for designers from other disciplines. The collection of practical case studies and tutorials will inform, help and inspire you to conduct research that will improve the processes, products, services and systems that you will design in the future.

1 WHAT IS PRODUCT DESIGN RESEARCH?

WHAT IS RESEARCH?

Research can be defined as the search for knowledge, or any systematic investigation into and study of materials and sources in order to establish facts and reach new conclusions. The word 'research' derives from the French *rechercher* (to search closely). The primary purpose of applied research (as opposed to basic research) is for discovering, interpreting and developing methods and systems for the advancement of human knowledge on a wide variety of scientific and humanitarian matters relating to our world.

Scientific research provides scientific information and theories to explain the nature and properties of the world around us. It makes practical applications possible. Scientific research can be subdivided into different classifications according to a multitude of academic and application disciplines.

WHAT IS DESIGN RESEARCH?

Unlike scientific research, design research is not concerned with what exists but with what ought to be. Research in a design context breaks with the determinisms of the past; it continually challenges, provokes and disrupts the status quo.

Whereas scientific research relies on and utilizes abstract mathematical explanations, design research uses representative images, narratives, physical models and 3D prototypes in the design and development of things that do not yet exist.

The process of design allows for the emergence of the unexpected and the intangible, and design research can be characterized by its willingness to embrace uncertainty, foster radical inquiry, and encourage ongoing dialogue with users, communities and the general public.

Fig. 1
Defankle Innovation Cards, developed as a tool to assist in the design process.

Design research also differs from scientific research in that it is concerned with the plausibility and appropriateness of proposals, while scientific research focuses on universal truths. Design research tends to produce knowledge that can be defined as trans-disciplinary and heterogeneous in nature, and that seeks to improve the world.

Investigating the process of designing in all its many fields, design research is related to design methods in general or for particular disciplines. A primary interpretation of design research is that it is concerned with undertaking research into the design process. Secondary interpretations refer to undertaking research within the process of design. The overall intention is to better understand and improve the processes, products, services and systems being designed. Design research has evolved into three distinct forms: research into and about design, research as design and research through design. Despite the inevitable overlap between these three categories, there are key distinctions.

1. Research into and about design (history, theory and context) is a common type of work undertaken in design research circles. There are many precedents for this type of scholarly work that looks closely at a specific area of practice. Such research, which is often historical, usually employs a method of critical investigation to evaluate and interpret a specific type of design practice and its significance. In this type of design research, the researcher is unlikely also to be the creator of the work in question. By reflecting theoretically on existing work from outside the creative process, the researcher is able to take a more objective view.

2. Research as design (innovative design methods) is a slightly more contested category. In recent years, a number of design researchers have claimed that making designed objects is, in itself, a process of research. This category refers to the notion that the outcomes of the research are, in some way, embodied entirely in the designed artefacts. The research is likely to involve the gathering and testing of ideas, materials and techniques required to make the artefacts. While research of this type is vital to the production of some original design work, it does not necessarily imply that the artefact makes an 'original contribution to knowledge' in the traditional sense of research. In this category the designer, who is the researcher, is operating almost entirely within the field of interest. The artefacts are unlikely to be interpreted from an external, objective position.

3. Research through design (experimental practice) is considered the taking of 'something' from outside the design work and translating it through the medium. Such work, commonly termed 'practice-based research', is often interdisciplinary in nature and can range from an idea or concept to a new material or process. In this case, the researcher will be engaged in making work within a field of interest as well as reflecting on it and contextualizing it. This reflective method gives rise to a viewpoint that is both internal and external to the subject of the research. In practice-based research new knowledge is generated by a combination of artefacts and the reflection that they engender. In this type of research the uniqueness and/or value will be contained in the nexus between the written text and the designed objects.

PRIMARY AND SECONDARY RESEARCH

Research can be divided into two forms: primary and secondary research. Secondary research involves the summary, collation and/or synthesis of existing research. Primary research involves the design researcher undertaking original research to collect new data through a range of methods and experiments. This enables the researcher to determine the size, nature,

Fig. 2
Digital technologies now provide designers with a vast array of computer-based tools to utilize in their research activities.

timeframe and ultimate goal of the research to be conducted. The drawback of conducting primary research is that the researcher has to create a detailed plan, and the time and cost taken to undertake primary research exceeds that of merely acquiring secondary research data and findings.

THE ITERATIVE DESIGN RESEARCH PROCESS

This book presents more than 50 design research methods, which have been categorized into seven essential phases, represented by the main chapters: Looking, Learning, Asking, Making (Prototyping), Testing, Evaluating and Selecting, and Communicating. These phases feed into each other through an iterative design process.

Opportunity identification
The earliest stage of the design process, often referred to as 'phase zero', begins with the identification of problems that need to be solved, needs that must be addressed and desires that need to be sated.

Brief and specification
This stage focuses on the construction and analysis of a design brief, identifying the customers' needs, and establishing a comprehensive product design specification (PDS).

Concept design
This stage of the design process involves the creation of a number of different viable concept designs.

Design development
This stage seeks to refine the chosen concept into a product that satisfies the requirements outlined in the PDS.

Detail design
This stage covers the key steps of transforming the chosen concept design into a fully detailed design, with all the dimensions and specifications necessary to make the product specified on a detailed drawing.

Production
The final stage involves how the product is manufactured, and focuses on determining what processes and techniques should be employed. It is important to remember, however, that some stages may occur in a different

sequence or may even be omitted altogether, as each product has its own unique set of requirements and the role and scope of design research may, as a result, vary.

As a design passes through the design process, it undergoes an iterative set of cycles, with each cycle consisting of five steps: understand, observe, visualize, review and implement. A cycle begins with the need to truly understand what research activities are required, enabling the design team to understand what needs and/or desires are being met or not. This is followed by a series of observations of end-users, to determine what is actually required. The visualization phase focuses on the production of a series of realized research outputs that enable potential or actual customers to engage with the concepts being developed and then critically analyse and review them with the design team. If the results of this feedback are satisfactory, then the design team can progress to the next stage of the design process. If not, then the design team can undertake the cycle again, prior to progressing.

PRODUCT DESIGN PROCESS AND METHODS

Typically, the creation of a new product begins with an idea and ends with the production of a physical artefact. Design research methods support the design and development of new products by providing invaluable data, expertise and knowledge through observing, recording and analysing how consumers interact with the designed world. Research has traditionally focused on the front end of the process, exploring the real needs and desires of end-users. Increasingly, designers are realizing the benefits of extensive research throughout the entire product design lifecycle, including investigating and evaluating the environmental, social and cultural impacts of a given product from its inception and manufacturing through its lifetime to its final disposal and recycling.

If you intend to minimize or avoid altogether any environmental impact, then you need to consider the effects of a product throughout its entire lifecycle: how it is produced, manufactured, transported, packaged, used and disposed of. The emergence of circular design principles has highlighted the need for products that are designed for longevity and recyclability, thus reducing waste and resource depletion. When designing new products, it is useful to storyboard the life story of the product to identify the possible impacts and events that may occur. This can be beneficial because the reality of many products' lives does not follow the exact route planned during the design research process, and so alternative eventualities should be considered. You need to establish exactly what the environmental impacts of the product are, or will be, and what they are caused by. By doing this, you can identify where the greatest need for improvement lies, and focus your design and research efforts effectively. Ignoring environmental factors in the design process means that designers are creating financial time bombs for their clients.

In addition to adopting a sustainable model of design, researchers need to follow an ethical approach to their research activities. Today's designers must also consider the role of emerging technologies, such as artificial intelligence and machine learning, in their processes, ensuring responsible usage and minimizing bias. You should fully reflect on the ethical implications of your work and how you conduct your research. Ask yourself: Who will own the data? Who can have access to it? Are you inadvertently exploiting the people you want to engage in the research? These questions are essential in the context of equity, diversity and inclusion principles, which call for a collaborative and respectful approach to research that acknowledges diverse perspectives. Creating products is a complicated process, and designers need to be aware of the larger contexts surrounding their work.

Chapter 1 What Is Product Design Research?

Fig. 3
Each iterative cycle includes five distinct stages, which are usually passed through before the designer either repeats the cycle to gather additional research data, or, if satisfied with the research undertaken, moves on to the next stage of the design process and the next cycle of design research and development.

Fig. 4
An illustrative example of research methods and tools and where they might be employed at different stages of an iterative design process.

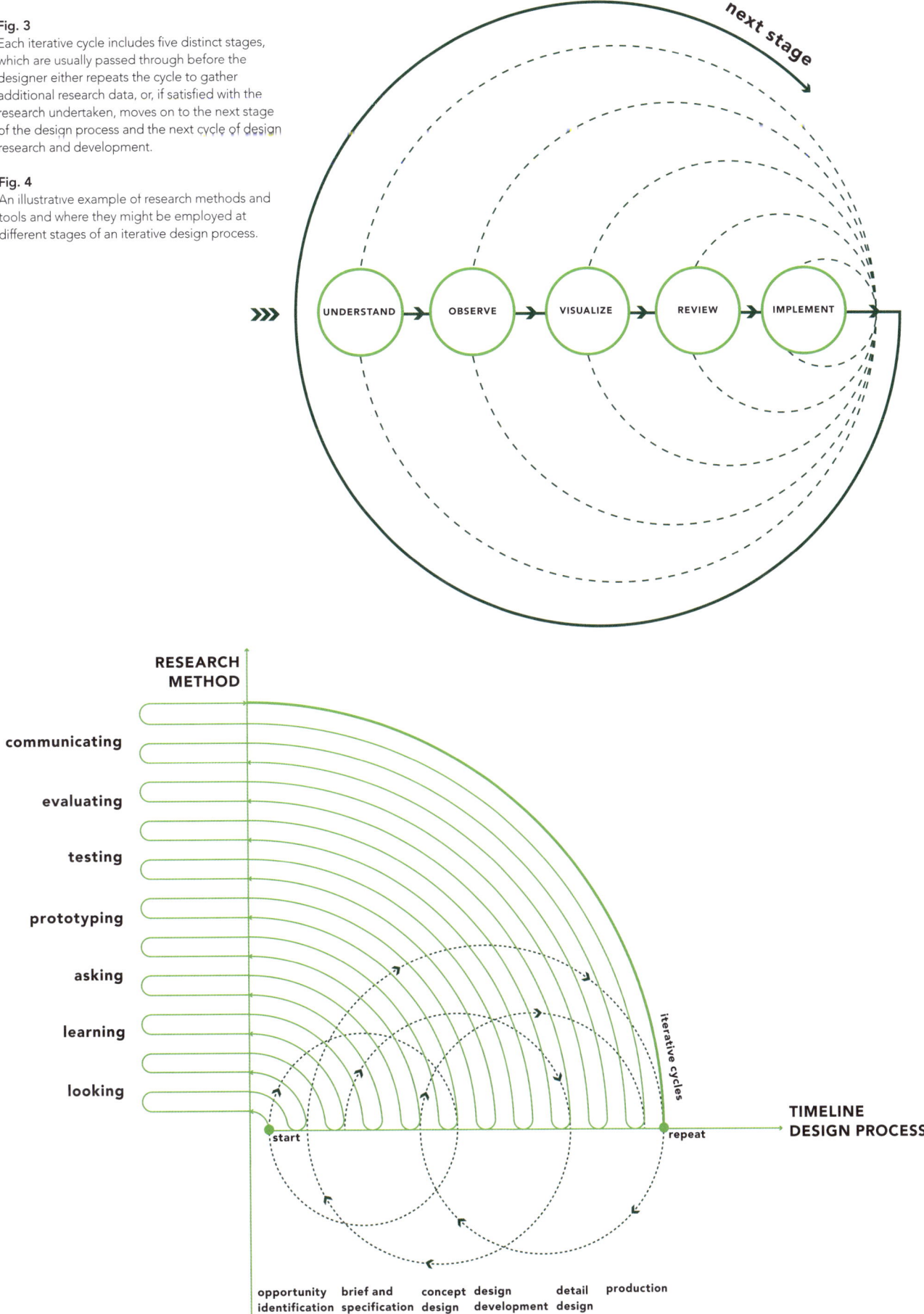

SUMMARY OF RESEARCH METHODS INVOLVED IN THE DESIGN PROCESS

This chart details the different stages of the design process and provides a list of the research tools and techniques that may be used, and where to find them in this book. The design process is more cyclical than linear (as shown on page 13) and you may be required to adopt techniques from both earlier and later chapters at every stage.

SCOPING

- Ethnography
- Photo & Video Diaries
- Shadowing
- A Day in the Life
- Personal Belongings
- Future Forecasting
- Product Autopsy
- Sketching

- Cultural Probes
- Competitor Product Analysis
- Literature Review
- Web Searches
- Cultural Comparisons
- Role Playing
- Try It Yourself
- Mind Mapping
- Sampling

- Questionnaires & Surveys
- Focus Groups
- Interviews
- Marketing & Retail Research

- Intuition
- Crowdsourcing

- Preparing A Presentation
- Report Creation

THE BRIEF

- Cultural Probes
- Competitor Product Analysis
- Literature Review
- Web Searches
- Cultural Comparisons
- Role Playing
- Try It Yourself
- Mind Mapping
- Sampling

- Ethnography
- Photo & Video Diaries
- Shadowing
- A Day in the Life
- Personal Belongings
- Future Forecasting
- Product Autopsy
- Sketching

- Questionnaires & Surveys
- Focus Groups
- Interviews
- Marketing & Retail Research

- Preparing a Presentation
- Report Creation

CONCEPT DESIGN

- Questionnaires & Surveys
- Focus Groups
- User Narrator
- Interviews
- Brand DNA Analysis
- Marketing & Retail Research
- Be Your Customer
- Image & Mood Boards
- Perceptual Mapping
- Personas
- Product Collage
- Extreme Users

- Sketching
- Photo & Video Diaries

- Cultural Probes
- Mind Mapping

- Intuition
- Matrix Evaluation

- Preparing a Presentation
- Report Creation

background stage
exploratory stage

DESIGN DEVELOPMENT

- Sketch Modelling
- Mock Ups
- Appearance Models
- Paper Prototyping
- Quick-&-Dirty Prototypes
- Experience Prototyping
- Empathy Tools
- Bodystorming
- Rapid Prototyping

- Photo & Video Diaries
- A Day in the Life
- Future Forecasting
- Product Autopsy

- Focus Groups
- User Narration
- Interviews
- Be Your Customer

- Intuition
- Matrix Evaluation

DETAIL DESIGN

- Scenario Testing
- User Trials
- Product Usability
- Material Testing
- Safety Testing

- Mock Ups
- Quick-&-Dirty Prototypes
- Experience Prototyping
- Rapid Prototyping

PRODUCTION

- Checklists
- External Decision Making
- Intuition
- Product Champion
- Crowdsourcing
- Matrix Evaluation

- Preparing a Presentation
- Report Creation
- Presentation Visuals & Models
- Engaging the Public

LEGEND

- CHAPTER TWO: **LOOKING**
- CHAPTER THREE: **LEARNING**
- CHAPTER FOUR: **ASKING**
- CHAPTER FIVE: **MAKING**
- CHAPTER SIX: **TESTING**
- CHAPTER SEVEN: **EVALUATING**
- CHAPTER EIGHT: **COMMUNICATING**

Fig. 5
Product design life cycle.

ANALYSING RESEARCH

The vast majority of research undertaken by designers is qualitative rather than quantitative. This requires the designer to develop a method of categorizing and analysing the data to uncover and understand the big picture, and highlight the important messages and findings to colleagues, clients and customers.

Once you have conducted a range of research activities and experiments, you need to step back from the methods and processes, and look for the themes, patterns and relationships that are emerging from your research. Look for similarities and differences, and see what different users are saying to you. You may find exceptions, contradictions and surprises, and through questioning, reviewing and analysing these you should be able to determine the key research findings that will inform your design process.

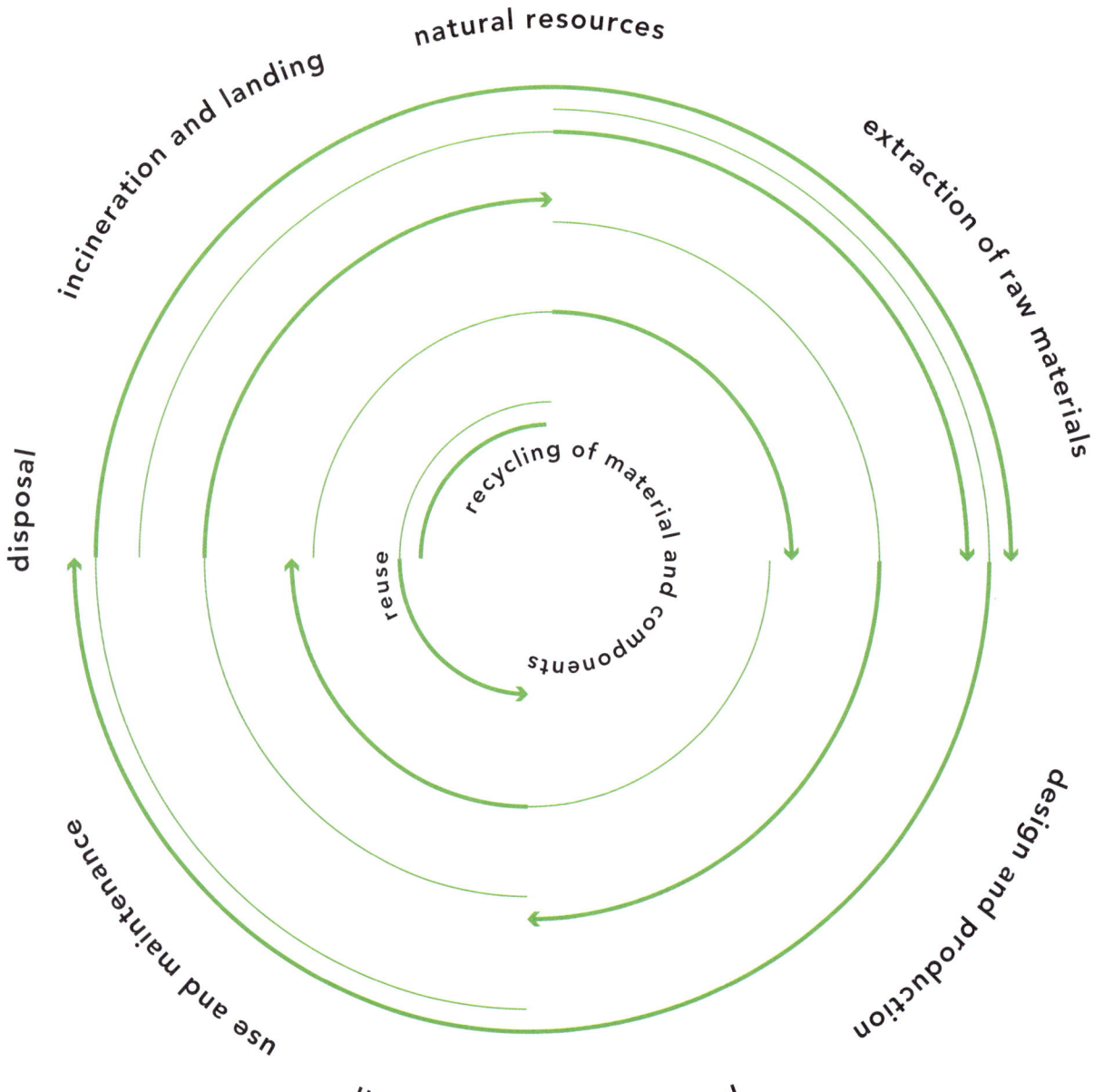

Whatever type of design research approach you plan, in common with other forms of research, the methods that you employ and the subsequent analysis of the research must be:

— Systematic
— Rigorous
— Critical
— Reflective
— Ethical
— Sustainable
— Communicable to others

ETHICS

Ethical design research requires carefully considered planning and preparation. Navigating ethical questions can be very challenging, and legal and ethical frameworks can often struggle to keep up with new approaches, such as AI and the increased use of data analytics.

When initiating a design research project, it is crucial to recognize the expertise and autonomy of participants in the process, and actively consider their needs and safety. Given that design practice and research increasingly engage people in co-design processes, it is imperative that you appreciate the inherent capabilities and creativity of your design research participants and the people you are designing with. By actively listening, clarifying your understanding and taking careful notes, you can do justice to the diverse stories and perspectives they share with you.

At the start of any research project, you must explain your research intentions to your subjects and participants: that is, what you are looking to find out and why. You need to describe clearly how you are going to use the information you collect and its value to you as a product designer. You should prioritize obtaining informed consent from participants and protecting their data, and you must gain permission from them if you intend to photograph or video them, and inform them that they can decline to answer specific questions or stop the research at any time. You must ensure that the data you collect remains confidential unless you have prior agreement with your participants.

Throughout the design process, you must be aware of any potential risks to those involved in the research process or impacted by it, and strive to prevent or address any harm that may arise. Transparency is essential, and communicating your goals and methods to participants enables them to ask questions and seek clarification throughout a project.

It is important to include marginalized and overlooked voices in your design research, but avoid tokenizing them or expecting them to represent an entire group. You should promote both inclusive and diverse participation, and wherever possible interview participants in their preferred languages while meeting them physically or virtually in safe and secure spaces. Make sure you treat people with courtesy at all times, and retain a consistently non-judgemental, relaxed and harmonious relationship throughout the process.

Design can be a high-paced industry, and it is important that you take care of yourself, while also ensuring that you are mindful of others involved in the design research process. Finally, all designers should commit to continuously discussing and refining their ethical practices in design research, encouraging themselves and their peers to reflect, ask challenging questions and engage critically with these ethical principles.

The following questions provide a simple ethical framework to adhere to throughout a design research project:

Design research planning and preparation

- Will your research process and findings account for human and ecological values?
- Are you actively accounting for direct and indirect stakeholders?
- Are you establishing realistic expectations for your design research and your colleagues, clients and research participants?
- Are your planned activities considerate of people's time, needs and expectations?
- When is the best moment to communicate your project objectives and the intended outcomes of participants' involvement?

Design research participant recruitment and engagement

- What perspectives are missing from your design research, and how can you connect with people who bring those viewpoints?
- Could a participant's involvement potentially pose future risk to them?
- Are there any potential conflicts of interest?
- Are your design research and recruitment strategies adequately accessible?
- Are you presenting yourself authentically?
- Are you prioritizing the unique needs of every research participant?
- Are you proactively verifying consent with participants during critical research moments?
- If any distressing issues arise during the design research process, are you consulting the wider research community and institutional/corporate guidelines, while ensuring confidentiality?

Gathering information

- Are you maintaining the integrity of your design research by focusing on listening rather than advising?
- Are you treating participants with respect by discussing them and their experiences as though they are present with us?
- Are you accurately quoting participants without misrepresenting their words or meanings?
- Are you taking all necessary steps to ensure the privacy of your design research participants and considering how their data will be used?

Using and sharing

- Are you acknowledging those who deserve recognition for their contributions?
- Have you done everything possible to improve the situation for the participants and communities you have engaged with?
- Whenever feasible, are you providing feedback to participants about the work you have accomplished?
- What strategies do you have in place for sharing results, and how will you ensure accountability for this?

Closing the loop

- What are your plans for sharing outcomes with participants?
- How will you ensure that you follow through on these commitments?

While these questions provide guidance for conducting ethical design research, it is important to recognize that at times it may not be appropriate to involve users and/or experts within a particular project. Designers should also fully acknowledge that the unpredictability of the design research and implementation processes can pose unexpected ethical issues that were not originally planned for, and therefore the support of ethical gatekeepers such as tutors or studio directors is always advised.

REFLECTING UPON YOUR RESEARCH JOURNEY

Reflection is fundamental to the design process, particularly within the context of research, as emphasized by philosopher Donald Schön's concept of 'reflective practice'. Schön argues that effective professionals engage in a continual cycle of reflection that informs their actions and decisions. This is essential in research, where reflection influences every stage, from selecting a field of study and the methods and tools you employ, to how you analyse your findings and determine subsequent iterative steps.

While designers often engage in reflection concerning their creative work and personal identities, it is crucial to delve into what reflection means specifically in a research context. Researchers are not neutral observers; instead, their perspectives are shaped by personal background, experience and training. Implicit biases can inform the interpretation of data, often unconsciously affecting outcomes.

To maintain the validity and integrity of your research, you should explore how your perspective or outlook has developed. It is also important to be aware of how the perspectives of your client, company and/or collaborators have evolved and how these relate to your research.

You should ask yourself what cultural, political or historical influences have shaped your personal beliefs and those commonly held. Can you ensure that your preconceptions do not shape or predetermine what and how you research? You could write a reflective statement on your positionality to gain insight into these questions, helping enrich both the research process and its outcomes.

2 LOOKING

In order to discover what people really need, want and do rather than just what they say they need, want and do, designers have developed a series of observational research methods. From the stories revealed through examining people's emotional attachments to their personal artefacts, to the manufacturing and aesthetic design decisions that are revealed through the forensic analysis of designed products, this chapter examines a number of the research methods used by leading design consultancies. You will discover, through carefully implementing these methods, how looking with a critical eye can help you design and develop great products to address the multitude of complex problems found in modern society.

ETHNOGRAPHY

Ethnography is a research method based on observing people in their natural environment – be that a hospital, a fish factory or a college library – rather than in a formal research setting. Ethnography is a very effective method for making sense of the complexities of people and cultures. It allows you to immerse yourself in other people's lives and witness patterns of behaviour within real-world contexts. Studying what people do rather than what they say they do provides a richer, more realistic overview of how people live, work and play. The aim of an ethnographic study is to learn from (rather than study) people in a particular cultural group – the intention being to understand those people's worldview. If conducted well, ethnography can provide you with a detailed, in-depth insight into people's behaviours, beliefs and preferences, and how they operate in their day-to-day lives.

In a design context ethnography helps designers create more compelling products, services, spaces and systems because they have been based on real observations of people.

Fig. 1
A sequence of photos highlighting the work of an automobile technician. Observing and recording visually what people do provides a rich and realistic snapshot of how people work.

The following points should help ensure you conduct a good ethnography:

1. Remember ethnography is not only about asking questions, but also listening to the answers.
2. Ethnography should delve deeply into the lives of a few people rather than study many people superficially.
3. Ethnography involves studying people's behaviours and experiences holistically in daily life.
4. Think carefully about what questions you will ask and how you will go about translating large amounts of data into concise and compelling findings.
5. Make good use of video, photos and other visual materials.
6. Avoid merely listing facts and focus instead on telling stories from the data you have collected.
7. Reflect carefully on the data you have collected and make connections.

Ethnography is a very powerful method and one a designer shouldn't dismiss lightly. If you want to design a product that will be accepted, used and loved by people, then you need to understand their day-to-day lives, rituals and habits first. Ethnography is a fluid process incorporating the collection, interpretation and presentation of data, which originates from the disciplines of social and cultural anthropology. Typically, in-depth, semi-structured interviews are employed in which respondents are encouraged to use their own vocabulary and participate as in a discussion. These interviews help uncover people's language, behaviour, thoughts and interactions with their products, environments and services.

Several variants of ethnography are commonly used in product design today. These include digital ethnography, where digital tools (e.g. cameras, laptops, the internet) can be used to speed up the process of data collection, analysis and presentation, and rapid ethnography, a method of estimation that fits well with the product design and development process, where designers tend to need answers in hours or days, not weeks, months or years.

Fig. 2
A visual array of research methods that can be used in product design (courtesy of IDEO Methods Cards).

Chapter 2 Looking

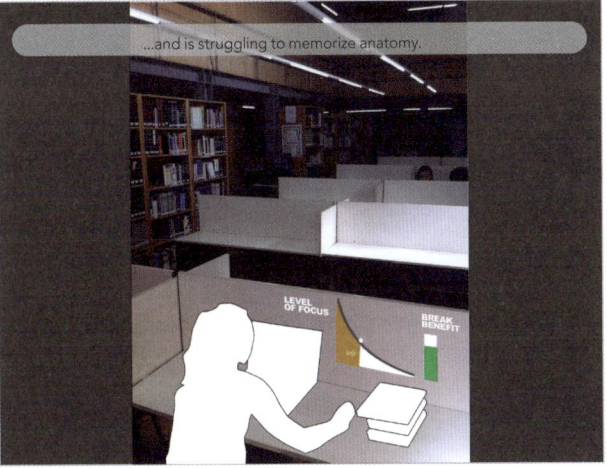

Fig. 3
Video diary recording the activity of a first-year medical student. This research was undertaken to determine the pace of life of students with a view to developing a 'widget' that gives feedback on their habits and guidance on how to have a more balanced lifestyle.

PHOTO AND VIDEO DIARIES

Photo and video diaries are a highly effective way of collecting observational, visually rich data. They allow design researchers to gain detailed insights into an individual's experience with a particular product or activity. A key advantage of photo and video diaries is that they help capture spontaneous and significant events and experiences in a person's natural surroundings, minimizing the time lapse between an actual experience and the person's account of the experience in their diary. Photo and video diaries provide an excellent way of studying and capturing significant moments in the day-to-day lives of people in natural settings, such as family birthday parties, breakfast, watching TV, reading the newspaper, cleaning dishes and bedtime rituals.

Research participants are asked to undertake a series of tasks on their own in their home or place of work and requested to photograph or film this as they do so. Each participant is typically given a set of instructions or prompts to follow. For example, participants might be asked to prepare a typical evening meal every day for one week and record their activity in a series of photographs, supported by written comments.

The participant would usually be asked to avoid taking photos of family members and friends during the study for confidentiality reasons. During a typical week-long study, the participant might be asked to carefully record the number, date and time of each photo, and where it was taken, together with any other comments they wish to make, such as why they chose that particular cooking utensil, why they chose to cook the food in that particular way, and so on. The participant would be asked to try to fill in the diary at least once a day for the full seven days.

Photo and video diaries give us access to a person's world beyond what we might be able to get in a face-to-face meeting, enabling us to see what they do in multiple locations, at different times and in a variety of situations.

Figs 4 & 5
Photo and video diaries enable researchers to collect valuable visual data, providing insights into an individual's experience with a particular product or activity over a set period of time.

Chapter 2 Looking

SHADOWING

Shadowing is a method that involves a researcher closely following an individual or small team within an organization over a predefined period of time. The researcher 'shadows' the target individual from the moment they begin their working day until the time when they leave for home. This can include hours of stationary observation while the person being shadowed writes at his or her desk, runs between buildings for a series of meetings or attends dinners held for clients. Shadowing is typically conducted over a number of consecutive or non-consecutive days for up to a month. Studies can be focused on a single role in several companies or on a number of roles within the same company.

Throughout the shadowing period the researcher asks questions that will prompt a running commentary from the person being shadowed. Some of the questions will be for clarification, such as what was said on the other end of a phone call, or the meaning of a departmental joke. Other questions will be intended to reveal purpose, such as why a particular line of argument was pursued in a meeting, or what the current operational priorities are. During shadowing the researcher will write an almost continuous set of field notes. They will record the times and content of conversations, the answers to the questions they asked and as much of the running commentary as possible. They will note the body language and moods of the person they are shadowing.

At the end of the shadowing period the researcher will have a rich, dense and comprehensive set of data, which gives a detailed, first-hand and multifaceted picture of the role, approach, philosophy and tasks of the subject. This research data can then be analysed in the same way as any other qualitative data, to provide invaluable insights, which can inform the creation of a detailed design brief.

When conducting a shadowing study you should always consider the following:

1. Never go in cold. It is important to spend time getting to know both the organization and, to a lesser extent, the individuals you will be shadowing. If you don't know the names of your subject's boss, work colleagues and husband, not to mention the major product lines and suppliers, your notes will not be very meaningful at the start of your shadowing.

2. Use a small, hardback notebook and a pen to keep a research account (and make sure you have a supply of spares). This will allow you to make notes wherever you are. Tape recorders are sometimes not practical for shadowing due to background noise.

3. Write down as much as you can. This is especially important at the start of a project when you can still see the organization as an outsider. Settings, the meaning of acronyms, how meetings make you feel, relationships and your first impressions of people (and how these change) are all relevant data.

4. Get into the habit of making a daily tape transcript of your research notes. This makes it easier to decipher what you have been writing at speed and helps keep your accounts rich and detailed. It also helps to preserve your own thoughts and impressions, which will change very quickly once you start to lose your beginner perspective.

5. Plan your data management. Decide how you are going to record, manage and analyse your data before going into the field.

A DAY IN THE LIFE

'A day in the life' is a useful research method for revealing unanticipated issues inherent in the routines and circumstances people may experience on a daily basis. It is an intensive research method that aims to provide a representative snapshot, in contrast to other observational research methods such as shadowing, which aim to build up a picture over a longer sustained period of study.

Designers using this method follow a 'performer' (someone who regularly performs the task) for a period of time, cataloguing the performer's activities and experiences throughout a typical day. The technique gets its name because the research is conducted over the course of a day. Typically, this will be for a full 24 hours or possibly a more conventional eight-hour, 9-to-5 work day. Using this technique, you act like a fly on the wall, observing and recording everything that the performer does, as well as the environment in which the performer interacts.

Fig. 6
Following an individual from the moment they begin their working day until the time when they leave can give a rich and comprehensive view of the role, approach, philosophy and tasks of the subject being shadowed.

For example, this might involve following and observing the work of an automobile technician in their garage or recording the daily activities undertaken by a dentist in their surgery. You might decide to record the activities in a handwritten journal and/or film the performer. In some instances, you may elect to follow an 'expert' and a 'novice' performer. By comparing their lives, you can understand how different levels of skill, aptitude and experience can affect how people navigate their world and their use of a range of products. You can then apply this knowledge to creating 'design interventions' that might help disadvantaged people.

In addition to observing the performer's behaviour and environment, you should also interview them throughout the day – at times that will not intrude on the work being observed. Interviews can provide insights into decisions that a performer makes and other behaviours that might not be visible or might not be clear from observing, as well as revealing the challenges that the performer faces on a day-to-day basis and the motivations underlying their work. When conducting 'A day in the life' research you should note down your personal impressions and any queries you may have that can be answered through subsequent primary research activities.

Fig. 7
'A day in the life' study, shadowing a nurse over the course of a full working day. The researcher acts like a fly on the wall, observing and recording everything that the subject does, as well as the environment with which she interacts.

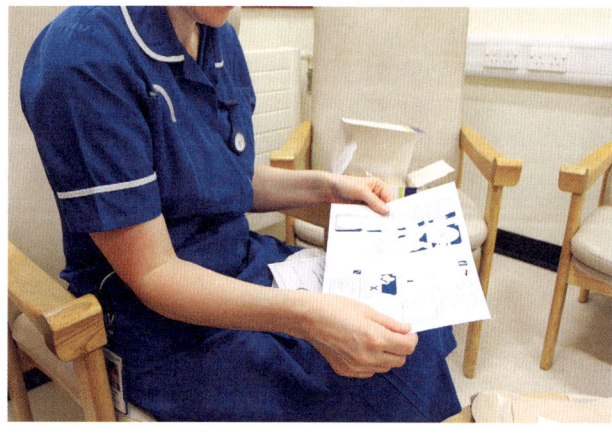
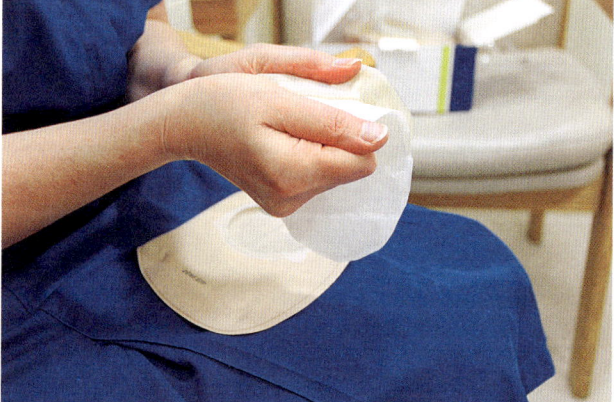
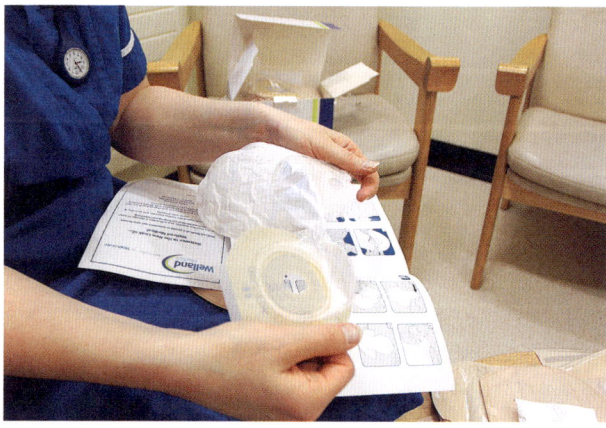
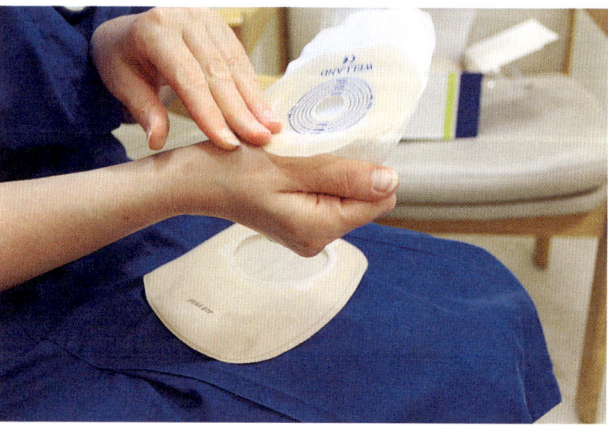

PERSONAL BELONGINGS

This research method involves asking people what personal belongings are important to them to ascertain the things that are valued by people and to identify any patterns of use among them. Owners and users often invest products with meanings that transcend their functionality, developing a patina of memories that can transform a seemingly mundane object into a treasured possession. This research method reveals how users assign personal meaning, symbolism and value to products. Information gathered through this method can then be used to inform the creation of products that address consumers' wants, rather than mere needs. In today's consumerist society, if the designs that we purchase, use and covet reflect who we are, then designers need to examine our relationship with our belongings. Through observation, designers can discover how people become emotionally attached to products, and apply these observations to create products that will be cherished. Valued personal belongings can be broadly categorized as follows, although they are not mutually exclusive.

Collectively cherished products

Iconic products and brands attract dedicated collectors who often form owners' clubs and fan sites to share mutual admiration and interest. These organizations and societies are fantastic repositories of research material for designers, and have been widely consulted when companies have chosen to develop 'retro' products that mine our collective memories, such as the electric Renault 5.

Individually cherished products

Everyday products can often gain emotional value through personal association and use. Heirlooms are products passed down from one generation to the

Fig. 8
Every Thing We Touch is a research method developed by Paula Zuccotti that documents daily life through the objects people interact with over a 24-hour period. By photographing and arranging these items chronologically, the method reveals personal habits, routines and cultural contexts – offering a visual ethnography that blends anthropology, design and storytelling. A day in the life of Mandy, 48, London, UK, by Paula Zuccotti, 2019.
everythingwetouch.org

next, and represent memories that transcend function or monetary value. Memorabilia and souvenirs always contain a built-in emotional value, such as the memory of a past holiday or significant personal event. Designers actively explore people's relations with their keepsakes, mementos and belongings to discover methods of creating products that foster meaningful relationships with their owners. Designers are often avid collectors themselves, and source obscure material to inspire their design practice. From the collection of vintage toys and mechanisms sourced in flea markets, through eBay, to priceless collections of design art and mid-century modern furniture, designers' personal belongings can offer an invaluable research resource.

FUTURE FORECASTING

Product designers need to continually strive to understand what their current and target customers want, how they currently use a product and what they will desire next season and beyond. Forecasting future trends is a key part of marketing and design strategies in the product design industry. Anticipating where the market will be in the future gives the designer an important research tool. While timely and regular market and research data can help to identify consumer needs, there is also a real need to anticipate future consumer wants, needs and desires.

Future forecasting can identify future aesthetic preferences, such as seasonal colours, materials and textures, through to developing speculative scenarios and products that aim to address long-term issues such as global warming. The gestation time to bring a product to market is such that the global conditions impacting on its success can alter from conception to product launch. With stockmarket crashes, socio-cultural trends and fashion moving so fast, expert research into future trends is becoming ever more important to the design industry and process.

Trend forecasters and futurologists (specialists who postulate possible futures by evaluating past and present trends) are commonly working to timescales of 18 months in advance in rapid-turnover fields such as fashion and textiles, to ten years or more in areas such as car design. Designers are required to analyse a wide range of subjective visual data and statistical information gathered from interviews with key stakeholders, reports from

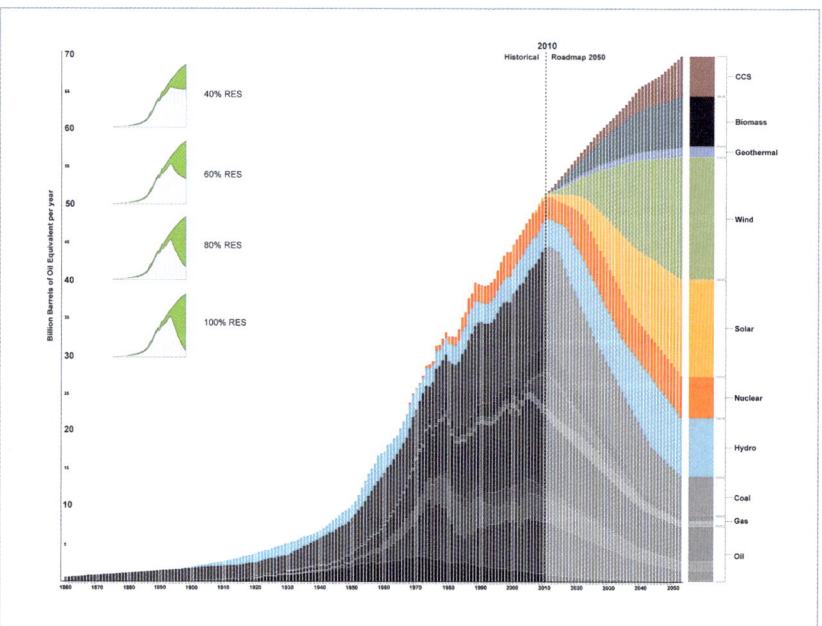

Fig. 9
The future forecasting method helps designers identify future aesthetic preferences, such as seasonal colours, materials and textures, and to develop speculative scenarios and products that aim to address long-term issues such as global warming.

photographic trend spotters, trade shows, exhibitions, selected key media and relevant social media. By distilling this information, they are able to identify emerging trends and extrapolate likely scenarios. These forecasts are then presented in the form of scenarios, stories and visual mood and image boards.

TREND SPOTTING

A trend by definition is something that has already begun, and trends are therefore spotted rather than created. Once spotted, a trend can then be analysed in detail, to identify its qualities and probable development. Trends are commonly presented through the following perspectives:

— **Commercial trends**: You can identify future economic trends and potential market developments and opportunities by critically reviewing sales figures, trade shows and market reports.
— **Design trends**: By collating a library of materials and samples you can predict design trends for identifiable periods. You can also use 'associational' words to summarize potential trends.
— **Visual trends**: You can create an informed overview of current trends and developments by making a visual survey culled from the design press, product launches, trade fairs, books, magazines and the output of numerous cutting-edge design groups.

Fig. 10
Trendsenses is one of a number of online forecasting services providing curated guidance on upcoming macro trends in design, materials and lifestyle.

Figs 11 & 12
Using scenarios to research and develop a range of digital musical instruments. Early research in this project studied how people reacted to various forms, and user scenarios were storyboarded to better understand the potential product use.

SCENARIOS, SPECULATIONS AND PREFERABLE FUTURES

Design scenarios allow designers not just to predict the future, but also to raise questions and issues about it, and to propose ideas and solutions that will enhance people's lives. Many product designers and companies now present scenarios through the medium of speculative prototypes. These speculative design concepts are communicated through narratives, exhibitions, images and films, allowing designers to propose design directions and garner subjective qualitative responses from a range of audiences. Future forecasting and trend spotting enable designers to predict and interpret the vital implications of user behaviour, and develop future scenarios informed by hard data and expert observation to give their clients the confidence that they are making the right decision.

Speculative design invites designers to challenge existing assumptions about technology, society and the environment. This methodology fosters a what-if mindset, encouraging the exploration of alternative futures that provoke discussion and critical thought about prevailing paradigms. By presenting design ideas that may not be immediately practical, speculative design opens up a dialogue about what could be possible under different circumstances.

Increasingly designers are acknowledging that their work has contributed to an unsustainable future by perpetuating existing systems of consumption and resource exploitation. Product designers can transcend mere problem-solving to actively challenge and disrupt these detrimental patterns, through

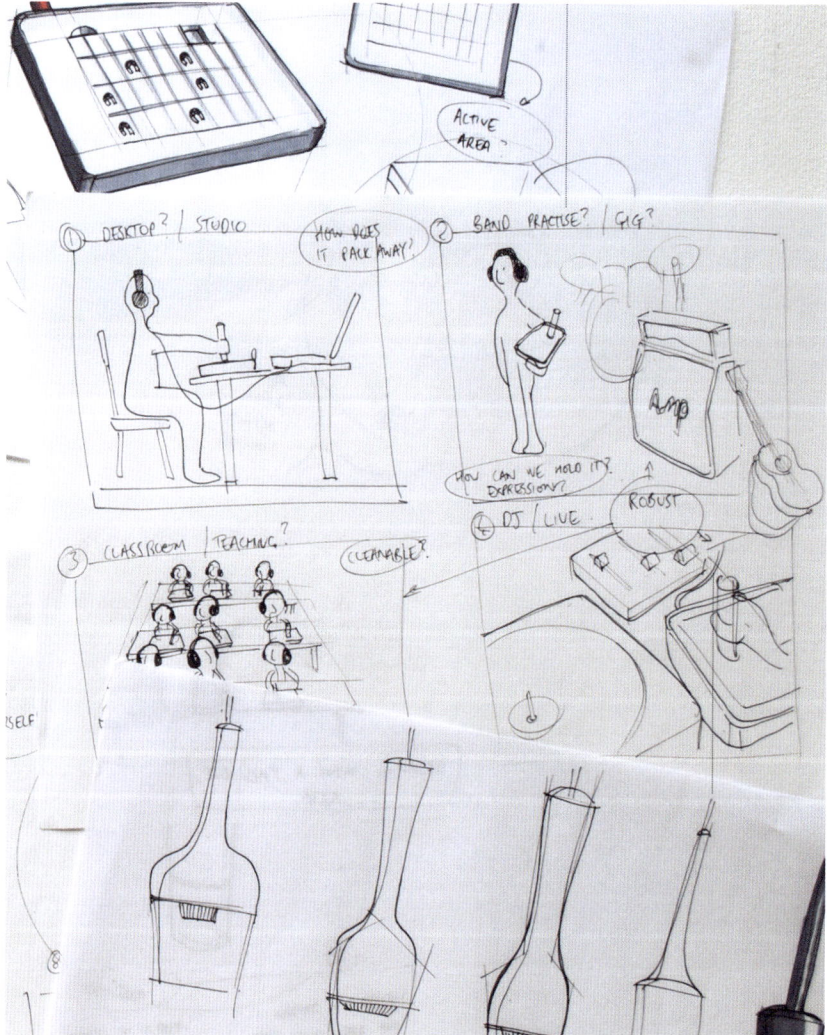

a process known as 'defuturing'. This entails reimagining design not only as a means of creating products but as a potent instrument for social transformation. By emphasizing sustainability, ethical responsibility and long-term viability, defuturing encourages designers to envision new possibilities that reject the disposability inherent in consumer culture and to embrace regenerative practice. Through this lens, design becomes a critical site for engaging with future scenarios, prompting reflection on the implications of design choices and their capacity to shape more equitable and sustainable futures.

Designers can tackle issues such as climate change and societal precarity by creating compelling scenarios that envision tangible alternatives, effectively communicating complex ideas and inspiring a sense of urgency regarding these pressing problems. For instance, designers might create a series of prototypes that visualize how sustainable practices can be integrated into everyday life, engaging people in a transformative vision of the future that prioritizes ecological balance. Scenarios could depict resilient communities codeveloping products in response to climate impacts or explore the ethical implications of emerging technologies, fostering inclusive dialogues around innovation.

When envisaging possible future scenarios, it can be helpful to refer to the 'Cone of Possibility'. This approach categorizes futures into the following types:

1. Possible futures: the full range of events that could unfold
2. Plausible futures: a set of possible but not likely scenarios
3. Probable futures: possibilities most likely to happen
4. Preferable futures: what you collectively want to happen

Designers and research participants could come up with several possibilities and scenarios in each of these categories.

By using speculative design methods, designers can craft concrete visions of future possibilities, encouraging audiences to reflect on their roles within these scenarios. This not only enhances engagement but also empowers individuals and communities to envision their agency in shaping preferable futures for both people and the planet.

Fig. 13
Joseph Voros' Cone of Possibility.

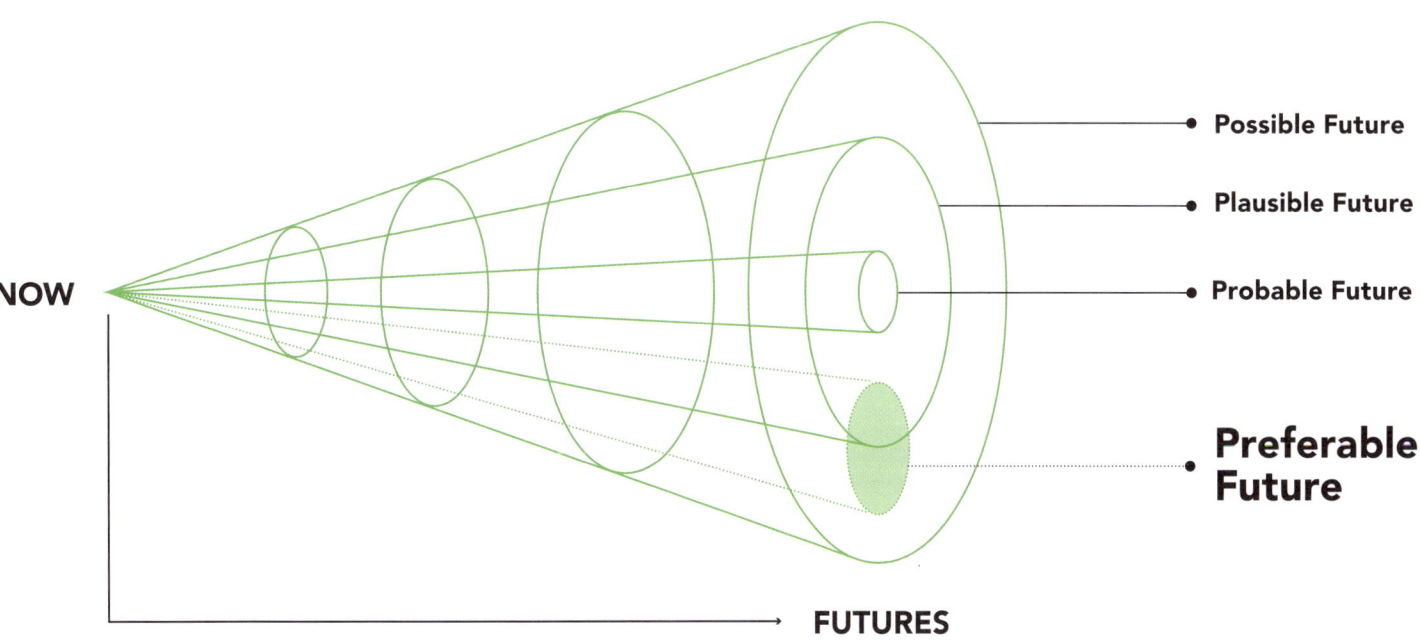

PRODUCT AUTOPSY

Product autopsy is a method used to gain a better understanding of the design decisions that have been made in an existing product, such as the materials used, the manufacturing techniques and processes employed in the product's development, why certain components have been used, and why aesthetic decisions on the product's form, colour and finish have been taken. A product autopsy is also a valuable method for reviewing how well a product has functioned during its life and how well it has aged.

When conducting a product autopsy, find out more about the product's life: how long it has lived, how well it has worked, which component parts have been damaged or worn and which have lasted. The autopsy begins with a rigorous visual analysis of the artefact, paying particular attention to any visible signs of wear and tear or damage. Next, the product is carefully disassembled. This will usually begin with the removal of a tough exterior casing to reveal many more fragile components and materials inside. It is typical, for example, for a mobile phone to consist of more than thirty major components, including parts such as an organic light-emitting diode display (OLED), electronics boards, a microphone, a speaker assembly, a display screen and battery.

Once inside the product, be gentle and thorough in finding out where the constituent parts have come from, how they might be reused or disposed of, how each constituent part was made, who made it, what its specific purpose is and how well it has fulfilled its role in the overall function of the product. Conducting a product autopsy enables you to make better decisions regarding material selections, spray coatings, electronics packaging, manufacturing processes and design choices for future products. As a designer you must remember that it is not merely a great aesthetic that makes a good product, but longevity and user experience, which win trust and ultimately build a loyal following. So your job as a designer is not only to design beautiful products, but also to ensure you design for the elegant integration of your products within a circular design system.

Fig. 14
Exploded view sketch by Gregor Whyte, exploring how a product is assembled and disassembled.

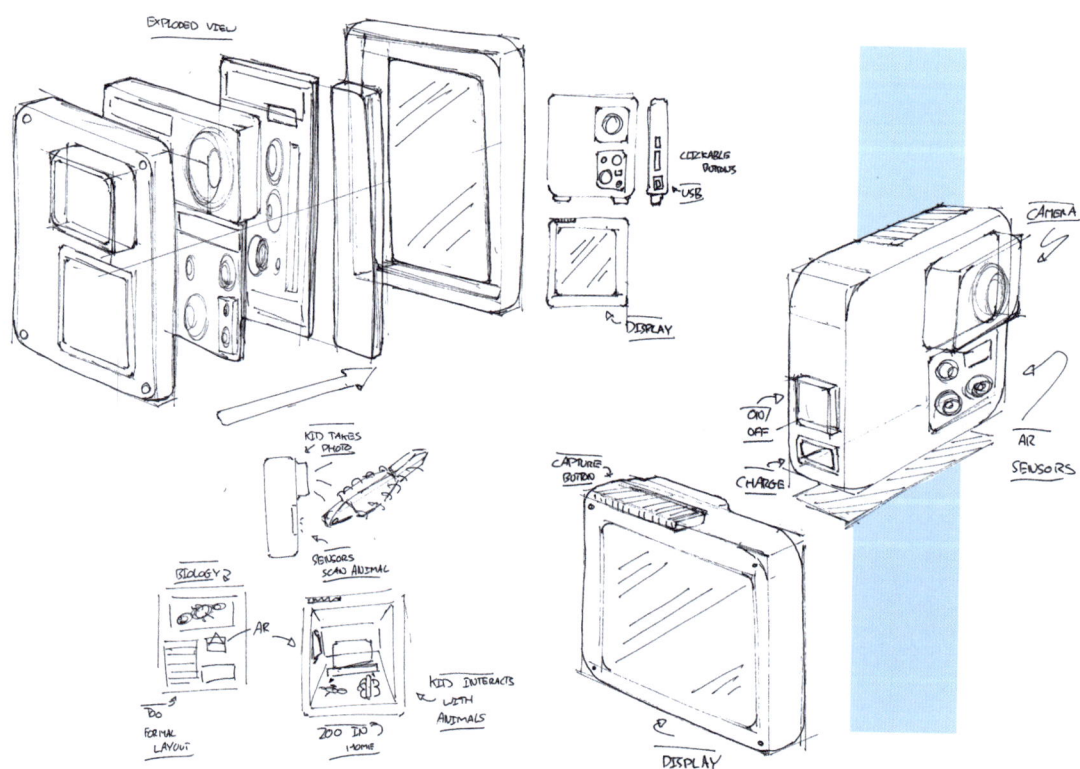

Chapter 2 Looking

Fig. 15
Carefully taking a product apart to conduct a 'product autopsy' allows researchers to better understand why certain design decisions were made, and allows them to see how a product's components have fared in terms of wear and tear.

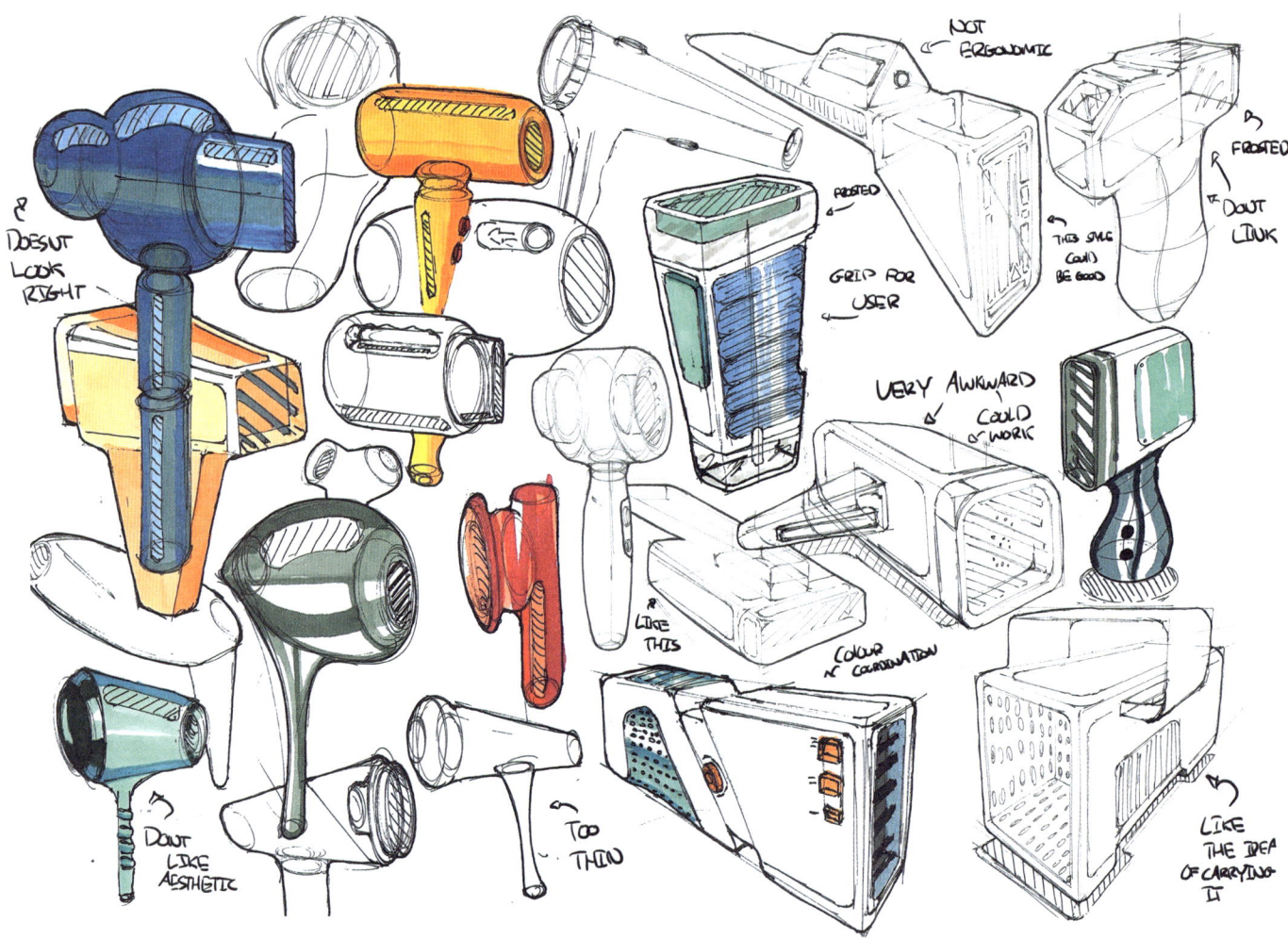

Fig. 16
Concept sketches by Gregor Whyte exploring form and function.

Fig. 17
Sketched plan for an exhibition at Italy's Triennale Design Museum.

SKETCHING

Sketching is a key research and development tool that enables designers to evaluate their ideas on paper, storing concepts for later discussion, manipulation and iterative development. The act of sketching is a means of firming up a research idea; it allows designers to wrestle with design possibilities, and attempt to give form and meaning to an idea.

Designers start generating their ideas with a pen or pencil and paper, or using a digital tablet. Most designers utilize these tools at the early stages of the design process because of the immediacy of the sketching technique, the freedom provided and the fact they can easily be erased, revised and redrawn. A designer also annotates his or her sketches – notes will act as aides-memoire for the designer and also help to identify key points so that his or her ideas can be communicated to members of the design team and the stakeholders involved. Concept sketches allow us to see the designer's mind at work. They fall into two broad categories, as follows.

Thematic sketches

These types of sketch are the initial exploratory visions of how a proposed design may look. They tend to be drawn in a wilfully fluid, dynamic and expressive manner, free from constraint. Thematic sketches should convey the product's physical form, characteristics and overall aesthetic. Such drawings often rely on a series of visual conventions that, to an uninformed eye or critical client, may need explanation.

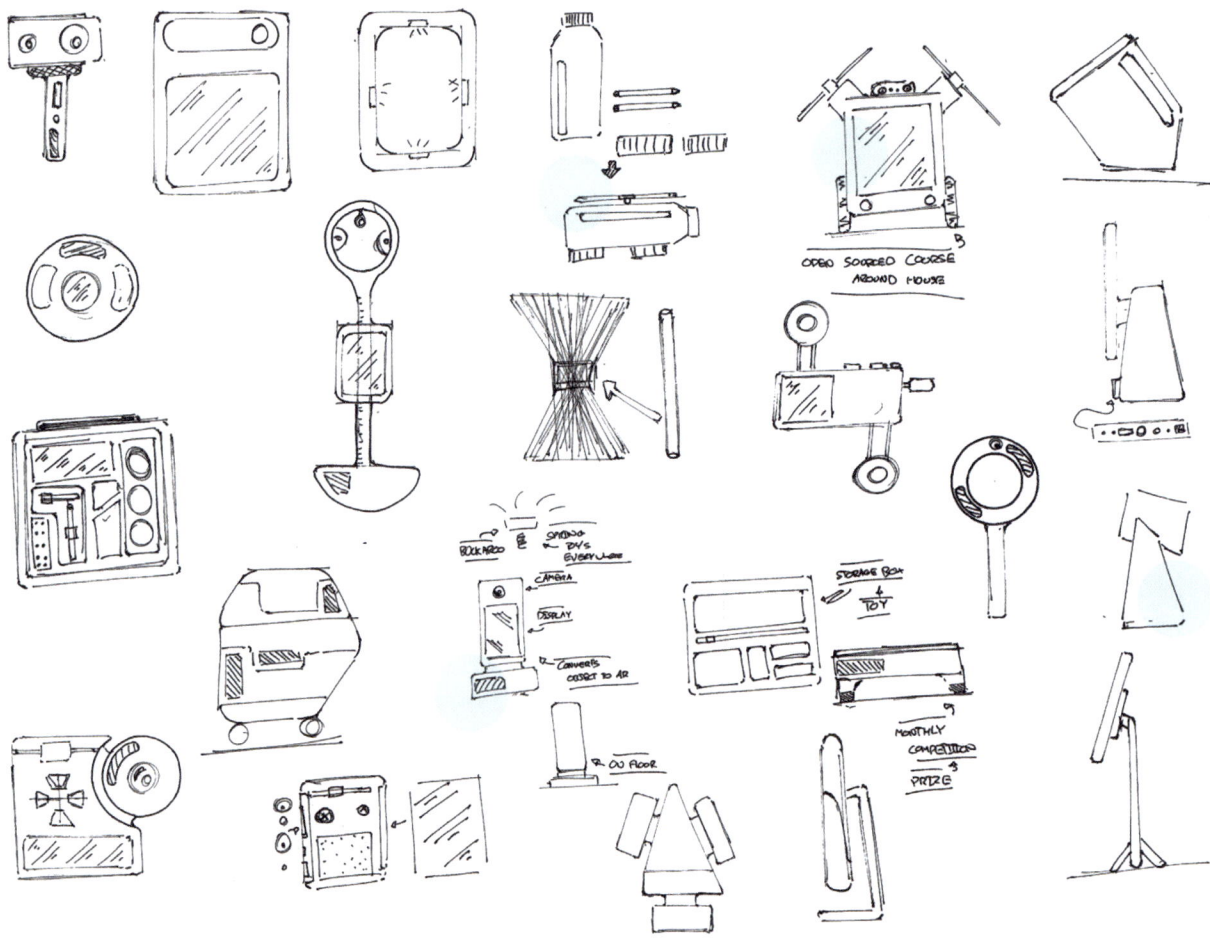

Fig. 18
(Above) Concept sketches by Gregor Whyte for WOVO, a creative device/camera for children.

Fig. 19
(Opposite) Detailed CAD drawings of Pearson Lloyd's Cobi Chair, with sketches and notes added to develop the backrest rib detail.

Schematic sketches

These sketches place less emphasis on the external styling or appearance of a design, and more on defining and working within a 'package'. This term is used to describe the fixed dimensional parameters of a design, including vital data such as off-the-shelf components to be used and ergonomic considerations.

The real secret to sketching is realizing that you should be as economical as possible with your marks on paper, while still creating an informative visual. Being able to sketch convincingly is vital for the designer, in order to communicate design concepts quickly and present concepts to users, colleagues and clients.

Once a concept has been settled upon, the designer is then able to move on to the production of presentation visuals to sell the design to clients or investors, and to involve target users in design critiques of these visuals.

Chapter 2 Looking

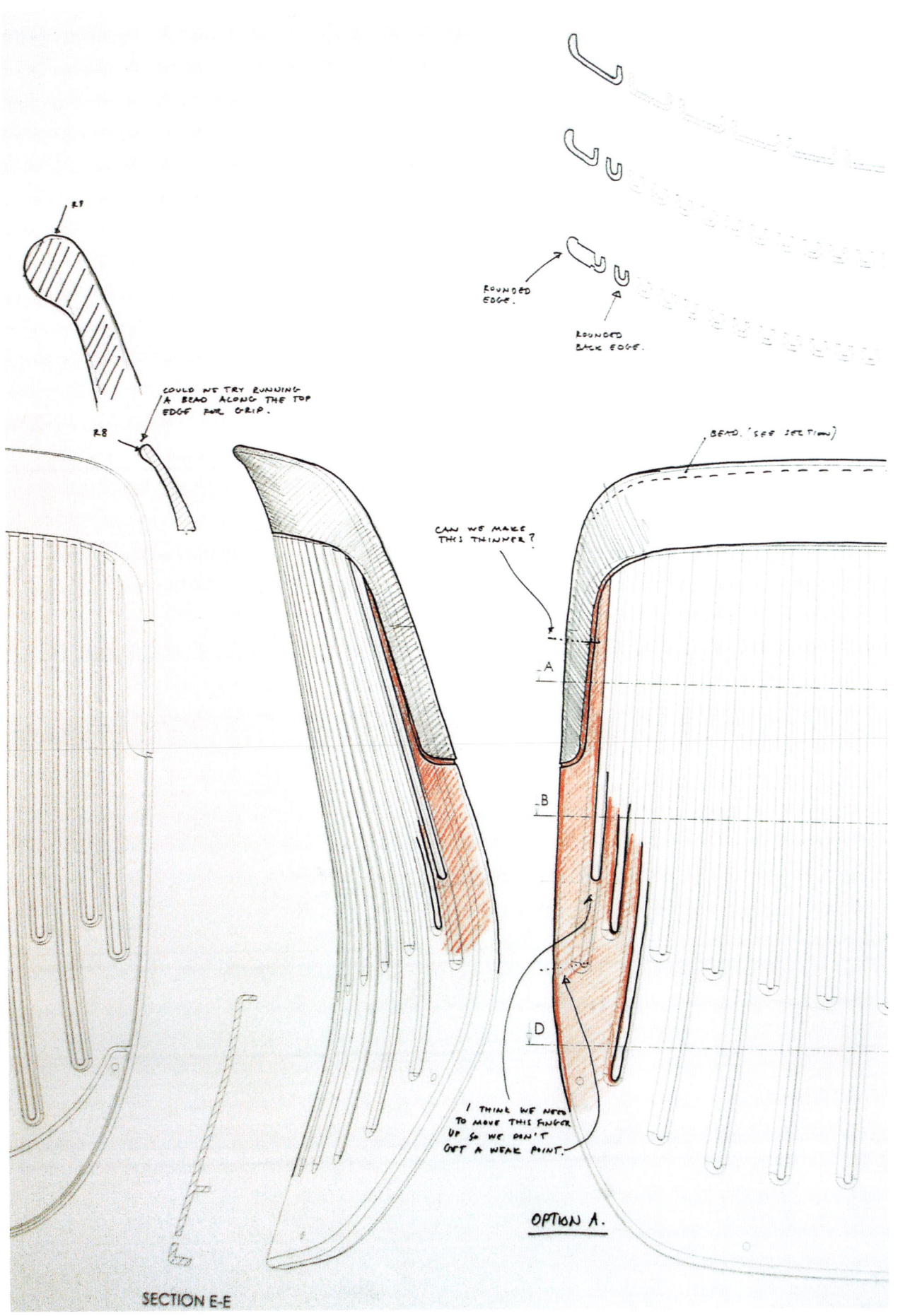

CASE STUDY

JINIL PARK'S DRAWING SERIES

Introduction
The core concept of Jinil's work is 'drawing'. The idea behind this project is quite simple: to bring furniture designs from sketches into reality. In an era flooded with countless design products, it has become increasingly difficult to create works that fully preserve the designer's original intent.

Objective
When Jinil entered university in 2007, industrial design related to products and automobiles was in vogue. The direction of all courses focused on sleek, modern design. After several years of study, he observed that during the process of transforming an initial sketch into a physical object, forms would inevitably change, and compromises were made to facilitate production.

Jinil grew frustrated with this aspect of design. Instead of adhering to modern aesthetics, he found inspiration in freely drawn, almost doodle-like sketches – ones that felt unrestrained, as if created without formal artistic training. This led him to embark on the process of bringing these sketches to life.

Methods
The 'Drawing Series' project explores elements found in the line drawings on paper. He realized that these lines could be transformed into compelling objects using steel wire. The key aspect of the work is the moment when a line becomes distorted. These distortions express the designer's emotions, state of mind and creative process. In design, a line plays a fundamental yet crucial role: it serves as the guiding element that defines both the start and end points of any piece.

From his furniture sketches, which originated from line drawings, Jinil selected the ones he found most effective and feasible to transform into solid objects. To achieve this, he used wires of varying thicknesses, hammering different surfaces with irregular force to shape them. This technique allowed him to replicate the spontaneity of hand-drawn lines in three-dimensional space. This process was the most time-consuming part of the entire project.

The processed wires were then welded at their intersections, ensuring structural stability. Through intuitive composition, Jinil combined these thin wires in a way that allowed them to support human weight – something a single wire alone could not achieve. This process successfully transformed two-dimensional (2D) drawings into three-dimensional (3D) objects, fully realizing the essence of his sketches.

Results
The 'Drawing Series' project exists at the intersection of design and art, between furniture and sculpture. This project brings freehand 2D sketches to life as tangible 3D objects within physical space. Unlike industrially designed products with sleek, polished finishes, Jinil's furniture pieces embrace wavy lines, subtle distortions in perspective, and exaggerated or omitted details. The final works appear light-hearted and stylish, yet the process follows traditional metalworking techniques, shaped by spontaneous intuition and the emotional state of the designer.
jinilpark.com

Opposite: Jinil Park has created a range of furniture from intersecting wires that has the appearance of a two-dimensional sketch.

CASE STUDY

PAULA ZUCCOTTI'S FUTURE ARCHAEOLOGY OF A GLOBAL LOCKDOWN

Introduction

Future Archaeology is a method developed by Paula Zuccotti to create time capsules of daily life through the physical traces we leave behind – our interactions with everyday objects. Just as archaeological artefacts have taught us how past civilizations lived, worked, played and expressed themselves, this approach asks: what stories will our objects tell about us?

It invites us to imagine what future generations might learn about our habits, needs and desires by studying the material culture of everyday life. Zuccotti first applied the method in her 2014 project-turned-book *Every Thing We Touch: A 24-Hour Inventory of Our Lives* (Viking, 2015). Combining art, design, anthropology and ethnographic research, the project documented people around the world by photographing everything they touched in a single day – arranged chronologically from morning to night.

The global lockdown in March 2020 due to the COVID-19 pandemic brought everyday life to a sudden halt. As routines and priorities shifted, these changes became visible in the objects people relied on. To document this transformation, Zuccotti extended the Future Archaeology method to lockdown life, inviting people worldwide to share 15 essential objects that helped them adapt to the new reality.

The resulting project captured a historic turning point – revealing how we worked, learned, cared, socialized, entertained, connected and loved during lockdown. Through objects that took on renewed or novel meaning, the project illuminated the transformation of our emotional and practical priorities.

Objective

'Future Archaeology of a Global Lockdown' set out to examine how the global crisis was reflected in our material culture, as shifting lifestyles transformed the everyday objects we depended on. Its central aim was to investigate which items became essential during lockdown and what they revealed about emotional, functional and cultural adjustment.

The guiding question – what are the 15 things helping you get through this time? – encouraged reflection on new items, rediscovered belongings or familiar objects with new significance. The goal was to create a global snapshot of adaptation and resilience through material traces.

More broadly, the project sought to evolve Future Archaeology as a method for documenting lived experience through everyday artefacts, and to consider how future generations might interpret this period through what we left behind.

Methods

To achieve this, Zuccotti launched a global participatory project via Instagram, using it as both a data-collection tool and an engagement platform. She began by posting her own 15 essential objects under the hashtag #EveryThingWeTouchCovidEssentialsX15, inviting others to respond.

Participants photographed 15 objects that had become central to their lives during lockdown. Each submission included a visual arrangement, a title and a short narrative. All images were produced in participants' homes – preserving intimacy and authenticity.

Submissions came from more than 50 countries, reflecting diverse ages, professions and cultural contexts. Zuccotti collected and organized the content entirely online, without leaving her home or using a camera. The project prioritized inclusivity, self-representation and emotional honesty. It encouraged participants to consider their relationship with objects in new ways

Chapter 2 Looking

– shifting the focus from aesthetics to personal meaning – and offered an effective tool for capturing the psychological and cultural dimensions of a shared global experience.

Results

The dataset – comprising more than 1,000 photo-narratives – was classified by country and city and analysed as a contemporary time capsule. Rather than idealized portrayals of lockdown, the submissions offered a raw and honest depiction of daily life.

The project revealed moments of personal growth and reflection. For many, lockdown became a time to slow down, reconnect with the self and revive old hobbies or interests. Some used the opportunity to develop creative skills or deepen their appreciation of simple rituals. These positive shifts – evident in objects like sketchbooks, instruments, gardening tools and journals – highlighted the potential for resilience and renewal amid uncertainty.

Participants also included items typically kept private: sleeping pills, cannabis, alcohol, antidepressants and books on mental health. These were shared without stigma, revealing how people coped with the emotional toll of the pandemic.

The photographs and stories reflected both individual adaptation and collective experience. Despite regional variation in object choices, recurring themes – care, connection, anxiety, comfort – transcended borders. While the pandemic affected people differently, it united them in a shared sense of vulnerability.

Although each image was consciously composed, the project resisted artifice. Its strength lay in its sincerity: each narrative, grounded in lived reality, offered a distinct yet universally resonant insight into the emotional landscape of a global crisis.

paulazuccotti.com
lockdownessentials.org

(Top) Fifteen essential items by Paula Zuccotti, London, April 2020.
(Above) Fifteen essential items by Sonia Skins, Taipei, April 2020.

TUTORIAL

How to conduct an ethnographic study

There are several important stages that you will need to consider and complete when undertaking an ethnographic study.

1. The first task involves defining the problem. You need to identify and define the key issues you are facing. You might have a specific question or just a general sense that more information is needed on a particular topic. Usually, however, the research question you ask will be fairly open-ended. Your objective is to observe, describe and interpret the situation you are studying. Your main research question might have several sub-questions that will be more specific in nature. Remember the key aim here is to examine the everyday activities and behaviours of people in their natural surroundings over a prolonged period of time.

2. The next stage requires that you identify and locate the people you wish to study. Who are the people who can most likely provide insights on the questions you are asking? Is it a particular group of people who use certain products or act in a particular fashion? Is it a particular cross-section of people that live and work in a specific environment, culture or geographical location?

3. Next you need to carefully plan your approach. You need to decide how you will actually carry out your observations and interactions with the people you wish to study. You also need to decide how much time will you allocate to each task, person and situation. You should develop a set of questions and prompts that will act as a consistent framework for your study. Make sure you also build enough time into the ethnographic study to allow your subjects opportunities to show what they do, what they value and how they undertake their day-to-day tasks.

4. The next stage involves the vitally important task of collecting data. This requires you to concentrate carefully on the task at hand, take everything in, use all of your five senses (not just your eyes and ears) and to be generally curious. The attitudes, mannerisms, language and interactions among the group you are studying are all very important. You must take a particular interest in how what you are observing might support or contradict commonly held assumptions. The collection of data at this stage will likely require you to take photographs, make video and audio recordings, and take handwritten notes and sketches.

5. The penultimate stage is the crucial task of reflection and analysis of data and interpreting opportunities. This is a demanding part of any ethnographic study. One of the key objectives in this stage is to go deeper than the more obvious insights derived from the observations and data collected in the previous stage. The analysis of ethnographic data can be a time-consuming activity. You need to be able to reflect carefully upon what you have experienced and observed during the study. This should always be done away from the members of the culture under study. The outcome of this stage might include design principles, films, personas and user scenario stories that can be told to multiple audiences, and might also include possible future steps.

6. The final stage involves the ethnographic research write-up and sharing of insights. Here, the insights that have been generated by the study are used to help inform the design decisions that will need to be made for the product being developed. This information should be presented in a highly visual way that will be much more likely to intrigue, inspire and engage your clients and colleagues, and can be communicated using techniques such as storytelling.

Observing people in their natural environment – in this instance a tree surgeon – provides a very good sense of the complexities of their day-to-day lives within real-world contexts.

TUTORIAL

How to conduct a day-in-the-life study

'A day-in-the-life' reveals unanticipated issues inherent in the daily routines and circumstances people encounter. In this example, a number of issues are identified in the way engineers install solar panels in residential properties.

Here are a number of tips to ensure that you get the most out of the 'day in the life' research method.

1. The first thing to consider before undertaking this type of study is to prepare beforehand. You should never start cold. It is vitally important that you spend some time getting to know both the environment of the organization where you will be carrying out the study and the person that you will be following closely. You need to identify the key individuals that will be integral to the study. For example, who is the line manager of the individual you are shadowing for a day? Who are the individual's key colleagues, and what are their names? What are the major work roles and activities of the person you are shadowing? What environments, interactions and relationships are crucial to this person in their day-to-day life?

2. Second, you need to keep meaningful notes. At times during the study your notes will be inarticulate and insignificant. This is fairly common and you may experience this at various points during the day. You should therefore plan how you are going to record, manage and analyse the data you wish to collect before going into the field. You might wish to structure the day around significant activities such as formal meetings, phone calls, presentations and so on. Alternatively, you might decide to break the day down into hourly slots. You might also restrict the time shadowing the individual to less than 24 hours – for example, you may shadow an office worker between the hours of 9am and 5pm.

3. Third, you must prepare the recording equipment that you will rely on to capture your data during the day. This will likely include a notebook and pen for jotting down notes and sketches, a camera for capturing significant moments in the person's day, and video and audio equipment for capturing important sequences and conversations. Any equipment will need to be compact; you must ensure that it doesn't get in the way of the person going about their normal daily activities.

4. Next, you must strive to capture as much data as you can – write down as much as possible, take lots of photos, and record significant moments using video and audio devices throughout the day. This is particularly important at the start of the day, when you can still observe the person, their organization and their work colleagues as a relative 'outsider'. The environments you experience during the day, the meaning of any work-related acronyms or industrial language, how meetings make you feel, the relationships among key workers and your first impressions of people (and how these change throughout the day) are all valuable forms of data and should be logged quickly and carefully.

5. When carrying out this kind of study it can also be helpful to discuss your research with a colleague or a friend. This individual should not be associated with the study in any way and should not be from the organization where the study is being carried out. Having an individual who is entirely neutral can provide vital moral support; they can also bring an entirely fresh perspective to your research.

6. Another good habit to get into is making regular data transcriptions of the research you are collecting. This makes it easier to make sense of what you have been writing at speed at particular points in the day and it also helps keep your accounts rich and detailed. Data dumps also help to preserve your thoughts and impressions at particular moments in time, which are subject to change as you lose your 'beginner' perspective over time.

7. Finally, you need to decide what to do with your data and how to make the best sense of it. A common approach for analysing qualitative data is content analysis, which usually involves a couple of key steps: (i) categorizing the information, identifying major themes and/or patterns and organizing them into coherent categories and (ii) identifying patterns and connections within and between categories. For example, categories might be theme-based (e.g. sleeping, walking, resting, working) or time-based (e.g. early morning, mid-morning, lunchtime, afternoon, early evening, late evening).

3 LEARNING

Designers can learn what people really need, want and do by looking closely at and learning from existing products, systems and services. Using careful and comprehensive information-gathering techniques, such as competitor product analysis, literature reviews and internet searches, designers can analyse and learn to identify patterns and insights into the behaviours of people and how they relate to and utilize their products. Furthermore, by employing role playing and research methods such as 'Try it yourself', designers can learn first-hand what using a particular product in a specific context feels like. A designer can learn a lot from the past by careful research, while at the same time keeping an eye on future developments.

CULTURAL PROBES

Cultural probes are designed to provoke, expose and capture the inspirational responses that describe an individual's relationship to designed products, spaces, systems and services. They are used to learn about people's lives as they watch television, search the internet for information, get ready for work in the morning, and communicate with family and friends, and so provide designers with rich data that can be used in the design of future products and services.

A cultural probe kit typically includes items for gathering a variety of information in a creative manner, such as an instant, disposable or digital camera, maps, sheets of paper, stickers, a diary, postcards, a voice recorder, pens and post-it notes. The contents of a kit depend largely on what kind of information you want to collect, and on the materials and equipment the participants are familiar with. People respond positively to good-looking, interesting kits, so careful planning, preparation and usage of good-quality materials is advisable.

Cultural probe kits can contain more than a dozen objects, each specifically selected to support the collection of myriad fragmentary insights into an individual's day-to-day life, wishes, needs and aspirations.

Fig. 1
An example of a cultural probe kit, a method first developed by Bill Gaver, Anthony Dunne and Elena Pacenti. Kits typically contain a camera, writing materials and a diary.

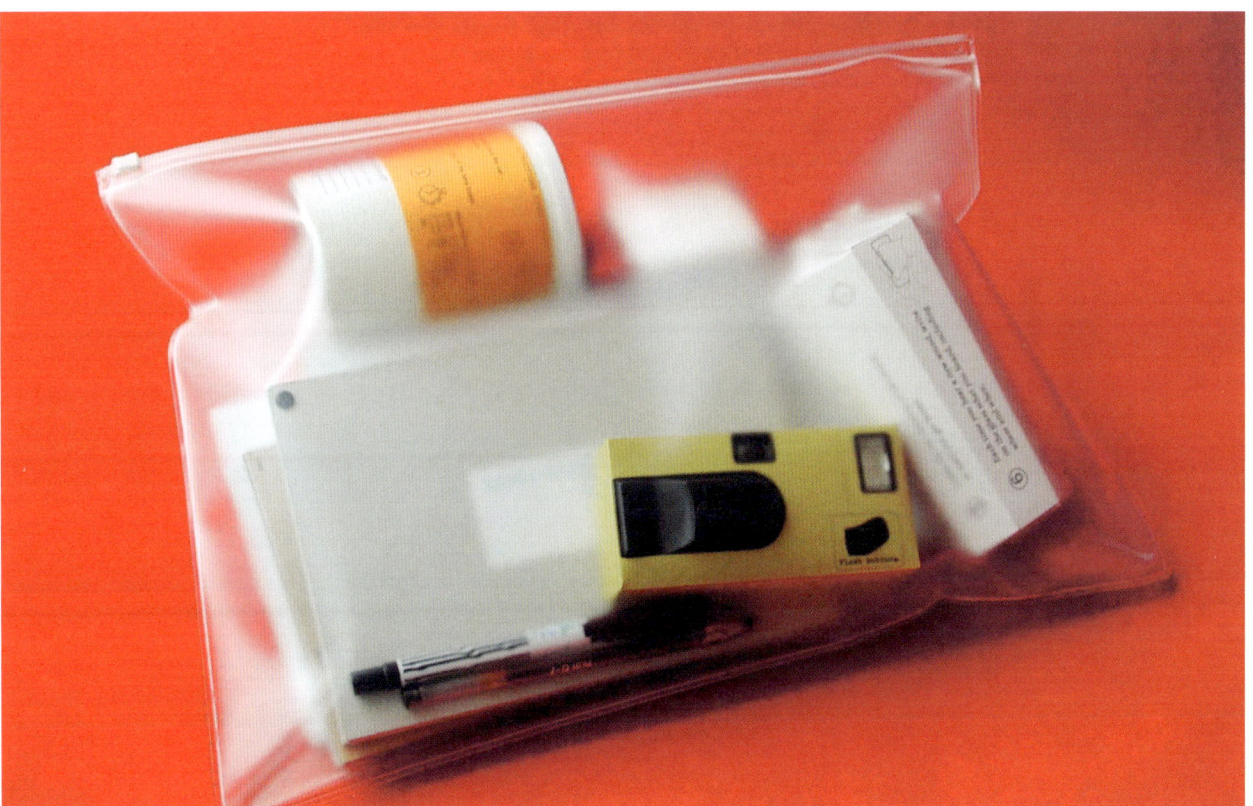

Using cultural probes involves a number of discrete stages, including:

— Planning what material (e.g. text, images) needs to be collected and how.
— Recruiting participants relevant to the design project.
— Selecting volunteers (e.g. teenagers, gardeners, teachers).
— Creating cultural probes (i.e. the objects and materials to be included in the pack), then deploying them – usually leaving the participant to complete the activities in the pack in their own time.
— Retrieving and analysing the probes.
— Designing the future product, service or system by processing the information collected in support of it.

Cultural probes are generally utilized at the concept-development stage of the design process to help support the definition of user needs and propose design ideas. One potential downside to using them, however, is that the amount of information processing and data analysis required can be time-intensive.

Cultural probes are particularly suited to design projects of an experimental nature – for example, where designers might be dealing with unfamiliar situations and need to understand local cultures and rituals so that their design proposals are not deemed inappropriate, irrelevant or arrogant. In short, cultural probes are intended to lead user groups towards unexpected ideas without pressurizing them into any specific single design proposal. Questions and prompts within a cultural probe kit typically pose open-ended questions and requests, such as:

— What is your favourite room in your home?
— What will you wear today?
— Use the camera enclosed to take six to ten pictures to tell us your story.
— Use the camera enclosed to take one picture of something boring.
— Tell us a place where you would like to go but can't.
— Tell us about your favourite product.

Researchers will deliberately word any questions obliquely, to provide users with as much room as possible to respond.

Fig. 2
ProbeTools are reusable digital cultural probes that provide invaluable insights and create a space of possibilities for learning about people, from the answers they give to questions, to pictures they take themselves, to animations recorded passively. Created by the Interaction Design Studio, the tools enable design researchers to gain access to what their participants get up to when the researcher is not around.

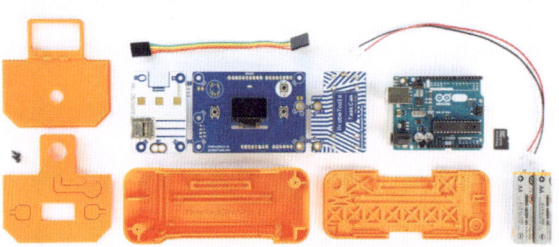
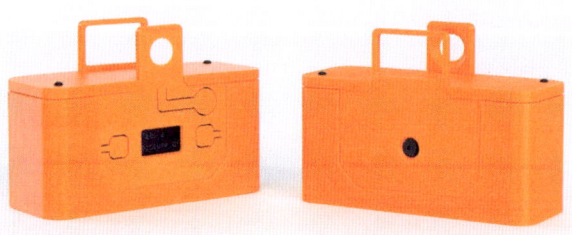

COMPETITOR PRODUCT ANALYSIS

Competitor product analysis is the process by which a product and its competitors in a specific market sector are examined and evaluated with respect to a predefined set of criteria. A competitor and/or analogous products analysis identifies the strengths and weaknesses of competing or similar products and/or services and is usually undertaken before starting to develop any design prototypes. This is a very useful way of evaluating and comparing the product being developed alongside its main competitors. A competitor product analysis helps to establish a range of both qualitative and quantitative criteria, including functional requirements and more subjective performance levels. Catalogues, trade magazines and the likes of the *Which?* magazine and website (formerly known as the Consumers' Association) are all good sources of details for competitor and analogous products.

Depending on the time and resources available, you begin the analysis by selecting an appropriate number of competing products to evaluate and compare. You should also determine exactly who the main competitors and their products are. (It is a good idea to ask a range of people, including users, designers, domain experts and marketing specialists, to review the list of competitor products to ensure that the most important competitors are represented.) Awareness of the product's intended use is vitally important, too, as it will identify the users, the tasks and the context in which the product will be used. A fundamental mistake in competitor product analysis is to focus too much on the technology and not on the user needs to be addressed. So you need to carefully define what products are already on the marketplace that satisfy the user needs you are interested in.

The competing products are then put through a series of typical tasks. By recording how each product performs during each task you will be able to evaluate each in turn by asking, for example: How easy (perhaps on a scale of 1 to 5) is the product to use? How reliable is it? How much does it cost? How long do I need to charge the product's batteries before they are fully charged?

Having established the strengths and weaknesses of each product, you will be able to generate a short summary of all of the competing products and their overall market position. This summary can then be used to develop a list of issues that need to be addressed in order to compete effectively in the marketplace. The summary may also help generate a list of desirable functions and/or features that the product being developed needs to include that were not considered at the outset of the project.

Fig. 3
A range of competitor products in the smartphone market.

Fig. 4
A design researcher reviewing relevant literature. Books, reports, conference papers, journal and trade magazine articles, and many other forms of output can provide an informed and critical account related to a particular design issue or area of exploration.

LITERATURE REVIEWS

A literature review examines published scholarly articles, papers, books, scientific reports and other relevant sources (academic dissertations, conference papers, trade magazine articles and so on) that provide an informed description, summary or critical account related to a particular design issue or area of exploration. The purpose of a literature review is to gain an overview of the significant literature published on a particular topic that will allow the design team to develop an informed opinion and perspective on the subject. It may, therefore, range widely in size and scope, depending on the nature of the design project being undertaken. A typical review may take in patent searches, legal reports, analogous product information, statistical data, government and private bodies' reports and market trends data.

A literature review has four main stages:

1. Issue(s)
what are the main issues under exploration? What are the relevant areas and what are the associated component issues?

2. Literature search
searching and locating published materials relevant to the issue(s) under exploration.

3. Literature evaluation
what are the key literature sources that will make a significant contribution to understanding and addressing the issue(s) at hand?

4. Analysis and interpretation
what are the major findings and conclusions of the literature that has been deemed relevant to the issue(s) under exploration?

It should be noted, however, that these stages may well be undertaken in an iterative fashion and will not always necessarily be completed in the order presented above. Careful attention should be placed on the evaluation, analysis and interpretation of each piece of literature, and full consideration given to the provenance of each article. What are the author's credentials? Are the author's arguments supported by evidence (e.g. historical material, primary research, case studies, statistics, scientific findings, etc.)? Each piece of literature under review should have a certain degree of objectivity. Is the author's perspective even-handed or prejudicial? Is opposing data presented or is certain relevant information ignored to strengthen the points the author is making in his or her paper? How persuasive is the author's writing? Which of the author's papers or reports are the most or least convincing? Finally, and significantly, what is the value of the literature that you have found and reviewed? Are the author's arguments and conclusions convincing? Does the work contribute in any meaningful and relevant way to the issue or exploration? Does it contribute a degree of knowledge and/or understanding to the subject you are exploring?

Although a literature review itself does not present any new, primary information, it is nevertheless a highly useful method for conducting research in product design projects.

CULTURAL OBSERVATION

Increasingly designers are observing how individuals and communities are autonomously designing. A good example of this can be found in how designers are drawing inspiration and research insights from such modification cultures as Ikea hackers and the lowrider community.

Ikea hackers think outside the flatpack, sharing their work to inspire others and help them in their quest for better products that meet their specific needs. Ikea has actively encouraged the hacker community by exhibiting examples within its corporate museum and website. Lowriders are cars that are customized to run low and slow, in contrast to hot rods that are customized for speed. Typically elaborately painted and decorated, often using graphic art of significance to Latino or African-American culture, this aesthetic is now influencing mainstream commercial design.

By observing cultures such as these, you can see the skill these individuals and communities have not only in designing products and modifying them, but also in the way they are designing new aesthetic languages, rituals of use and democratic forms of design.

INTERNET SEARCHES AND USING GENERATIVE AI AS A RESEARCH TOOL

The internet has transformed society, putting vast amounts of data at our fingertips and revolutionizing the way we conduct research. Search engines and generative AI platforms offer the ability to perform targeted searches for specific information by inputting relevant keywords or prompts, which are then processed against a vast database of web pages. However, the sheer volume of results returned by search engines can often feel overwhelming. To manage this, it is crucial to use specific keywords that accurately encapsulate your research query. Even if you receive several thousand hits, this overload can be navigated, as search engines typically rank results by relevance based on your input. It is also important to note that different search engines may produce varying results and relevance rankings; some may analyse the full text of a web page, while others may only consider the initial sentences. Therefore, utilizing multiple search engines allows you to cross-reference findings and broaden your dataset, assisting in evaluating the overall value of the information obtained.

Fig. 5
Websites such as the Arts and Humanities Research Council-funded Design Research for Change project provide invaluable peer-reviewed research papers, data and insights.

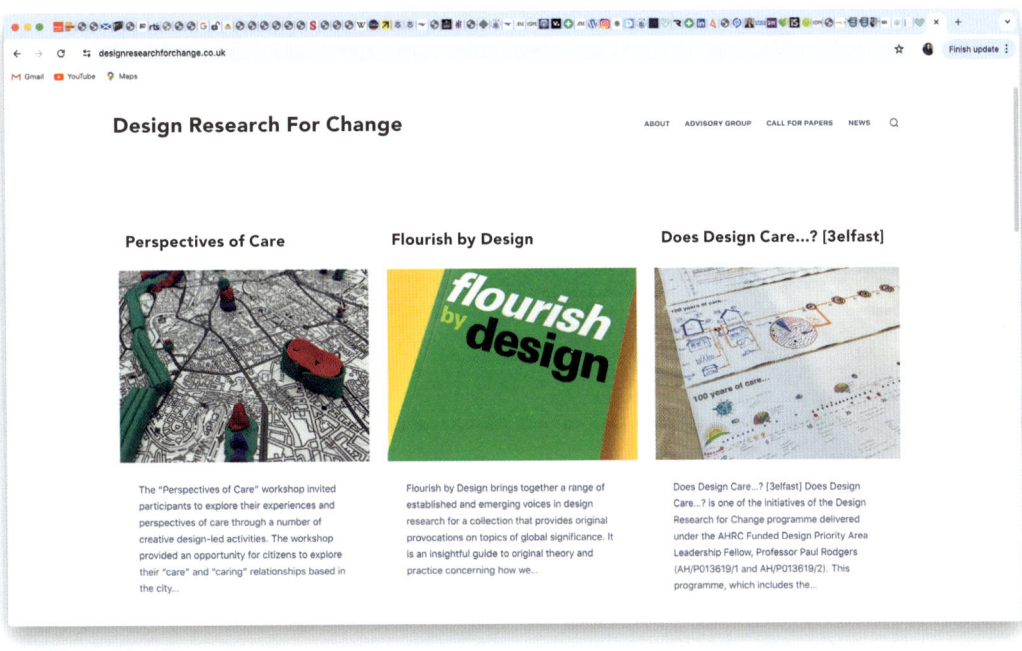

Fig. 6
It is a mistake to assume cultural groups are the same worldwide; the Western idea of the gloomy black-clad teenager doesn't necessarily translate to Japan, with its variety of subcultural groups who adapt the garish colours and styles of Japanese manga and anime.

Once you identify potentially relevant information, it is essential not to accept it at face value. Assessing the authority and credibility of the source is a vital step in ensuring a robust research process. Make sure to thoroughly evaluate the accuracy, reliability and objectivity of the data by considering the following questions:

— Authorship: does the information have a stated author or creator, whether personal or corporate, or is it presented anonymously? Understanding the source of the information can provide insight into its credibility.
— Background: what are the author's educational or occupational qualifications? Are they affiliated with a respected professional association, academic institution or credible publisher, or are they an unknown individual lacking recognizable credentials?
— Accuracy: does the information appear factual and consistent with other reliable sources? Cross-verification with multiple reputable sources can help confirm this.
— Review process: has the information undergone peer and editorial review, which can enhance its reliability and objectivity? Academic articles and publications typically adhere to this rigorous process.
— Currency: when was the information last updated? Always look for a publication or revision date at the bottom of the page and check that any hyperlinks to other resources are still functional and relevant.
— Objectivity and bias: how impartial is the information? Consider whether the data presents multiple perspectives or whether it seems skewed in favour of a particular viewpoint. Identifying any promotional intent, such as whether the content serves as commercial advertising or attempts to sell a product, can also reveal bias.
— Comprehensiveness: evaluate the depth and breadth of the information. Is it thorough enough to meet your research needs, or is it overly superficial?
— Target audience: consider who the information is aimed at. Does the audience level (layperson or specialized) align with your research requirements?

Chapter 3 Learning

By applying these evaluative questions to the information gathered, you enhance the robustness of your research process, ensuring that findings are based on credible, high-quality data that can effectively support your inquiries or arguments. Moreover, incorporating diverse forms of data, such as academic journals, reports and expert interviews, alongside internet searches can further reinforce your research by providing a richer, robust and rigorous understanding of the topic at hand.

AI LANGUAGE MODELS AS RESEARCH ASSISTANT

Artificial intelligence language models, such as ChatGPT, Claude and DeepSeek, can be useful tools for conducting design research as they allow you to engage in a conversational format that mimics an interactive dialogue with a knowledgeable assistant. To maximize this approach, start by formulating clear and specific prompts that convey your research questions or topics of interest. For example, instead of a broad instruction like 'Tell me about product design', refine your prompt to something like, 'What are the key stages involved in the product design development process?'

Additionally, use follow-up prompts to delve deeper into specific aspects of your research. For instance, after receiving initial information, you might ask, 'Can you explain how user feedback influences product design iterations?' This iterative questioning process helps clarify complex subjects and uncovers nuanced insights.

Moreover, consider prompting for various perspectives by asking questions like, 'What are the main challenges in sustainable product design?' This can help you to gather a balanced view on important topics within product development. Lastly, using prompts to request summaries of larger texts or explanations of specific terms enhances your understanding while saving you time. By strategically using AI in your research process, you can efficiently gather, clarify and synthesize information tailored to your inquiry.

While AI language models represent a significant advancement in artificial intelligence, they are not without notable shortcomings. One major limitation is their reliance on patterns learnt from vast datasets, which can result in inaccuracies or biases reflective of the training data. Consequently, they may produce information that is misleading, outdated or factually incorrect. AI language models often lack a deep understanding of context, leading to responses that may seem coherent but miss critical nuances or fail to grasp the specific intent behind a user's query. They are also unable to verify real-time information or access the internet for the latest updates, which can limit their relevance in rapidly evolving topics.

Fig. 7
An example of role playing a situation in a hospital. Role playing is an effective way of gaining a deep understanding of particular situations and of raising important issues from different perspectives such as the clients, the manufacturers and specialist end-users.

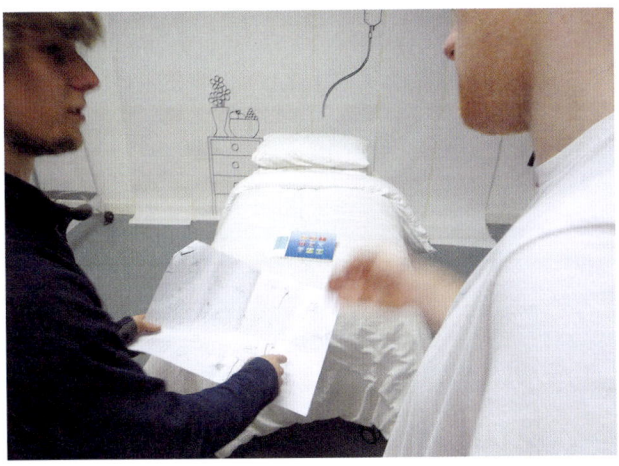

It is important to remember that AI language models do not possess true understanding, consciousness or emotions, which can result in responses that lack empathy or depth. Therefore, you should always approach outputs from AI language models critically, and ensure you validate AI-generated information through reliable sources while recognizing the inherent limitations of machine-generated information.

CULTURAL COMPARISONS

Greater societal awareness of diversity, equality and inclusion has compelled product designers to address an increasingly broad range of users and communities, necessitating a nuanced understanding of the specific needs of these varied users. In contemporary product development, designers must navigate distinct cultural sensibilities and communication styles across different countries and linguistic backgrounds. Cultural comparisons, as a research method, utilize both personal experiences and published accounts to reveal differences in behaviours and artefacts across cultural groups. This forensic approach is essential for understanding the cultural factors that impact design decisions, particularly when venturing into unfamiliar or global markets.

Manufacturers increasingly tailor their products to align with the cultural preferences and requirements of different international markets. These variations can encompass various elements, including branding, colour schemes and product formats. A pertinent example is Coca-Cola's approach to marketing its beverages differently around the world. In Japan, Coca-Cola introduced unique flavours, such as green tea and peach, to cater to local tastes, whereas in Europe and North America, the focus remains primarily on traditional cola flavours. These adaptations underscore that products designed for one region may require significant modification in another.

In the realm of technology, the mobile phone industry illustrates the importance of cultural sensitivity in design. For instance, Apple has recognized the preference among users in certain markets for smaller devices that facilitate one-handed operation. Research indicated that in high-density urban environments like Tokyo, where commuters often navigate crowded public transport, the ability to operate a smartphone with one hand is paramount. This has led Apple to design phones that accommodate this need, ensuring that essential features remain accessible with a thumb.

Moreover, the rise of niche subcultures, often referred to as 'urban tribes', has become increasingly influential in design. Members of these subcultures, such as skaters or K-Pop fans, express their identities through distinctive styles, often adopting specific products and brands that resonate with their group. Designers use semiotic analysis, a technique rooted in the work of philosopher Roland Barthes, to decipher the cultural codes associated with these subcultures, allowing them to create products that connect meaningfully with the intended audience. For example, Nike's collaborations with streetwear designers not only cater to fashion trends but also embody the values and aspirations of urban youth culture. By employing cultural comparisons, designers can create products that not only recognize but also reflect and respond to the diverse values, beliefs and expectations of different audiences and cultures, fostering a more inclusive and empathetic approach to design in an interconnected world.

ROLE PLAYING

Role playing is a very flexible and effective way of gaining a better understanding of the key stakeholders involved in a project, and of raising important issues. By adopting roles such as clients, manufacturers and specialist end-users, designers can simulate the real-life experiences and activities that are demanded in particular contexts. For example, designers might role play specific situations, such as driving a taxi, serving cocktails in a bar or changing a wheel in a garage.

Role playing requires you to project future scenarios – a kind of 'what if'. In any particular future scenario, you are projecting yourself and others into an imaginary situation where you cannot completely control or predict the outcome, but you can anticipate some or all of the conditions and 'rehearse' your performance in order to shape the discussions and outcomes.

By adopting the roles of others, designers can gain valuable insights into what motivates certain stakeholders and what their objectives and values might be. Role play is defined generally as an experience around a specific situation that contains two or more different viewpoints or perspectives. The role play situation can be organized around a prepared brief and the different role perspectives on the same situation are handed out to the different people who will come together to discuss it. Each role has a particular objective, or set of objectives, the person wishes to achieve, which may well be in conflict with other role players.

Role playing situations need to be realistic and relevant to the individuals playing the key roles; the most successful experiences focus carefully on developing particular situations and gaining knowledge and information from them. It is important to remember that role playing should be fun – it should not be a stressful experience for the people involved. It can help build design teams, develop employee motivation, improve communications and support excellent new product design and development. You should follow these simple guidelines to ensure you get the most out of role playing:

— Adequate preparation time is required – this might seem obvious, but it is often overlooked in the belief that it is best to get on with the role play quickly.

— Role playing must be focused – you must be clear at the outset about what you want people to achieve during the experience. Uncertain thinking at the outset will result in fuzzy outcomes. Clear thinking and preparation will result in clear and important outcomes.

— Role playing objectives must be clear and understood – if people are unclear about what they are supposed to do, your role play will be ineffective. The brief for everyone involved should be unambiguous: there should be clearly set objectives and enough information for people to engage in a believable and relevant conversation in line with these objectives. Be clear about the purpose of the role play situation, too. What information do you want to gain from conducting the session?

— Rehearse your role play – you will likely need to do it again and again to get it right. So you will need to rehearse again and again to get your behaviours and the role play relationships just right to make sense of the scene and understand the issues involved.

— Role playing ambitions – you need to be realistic in your ambitions for the role play. If you set your ambitions too high then you might cause people to lose confidence in themselves and in the role playing situation at hand. If you don't have time to get the participants to complete the whole exercise properly, in depth, with plenty of rehearsal and revisiting, then you may be better off completing just part of the situation.

Fig. 8
By putting yourself in the user's shoes, you can develop a better understanding of the issues that your customers will face, rather than merely looking at the product from a design perspective.

— Role playing feedback – this is very important for achieving successful situations. Not only can the participants provide feedback on the roles they are playing, but they can also benefit from feedback from observers. Observers should explain their feedback clearly, following SMART principles (Specific, Measurable, Achievable, Relevant and Time-bound). Observers should describe the specific things that they saw and heard, relevant to the exercise and to the people doing the role playing. Role play feedback should avoid subjective judgements based on personal knowledge or assumptions. It should also provide meaningful and specific comments that the role players can act on.

TRY IT YOURSELF

'Try it yourself' is a commonly used research method that enables designers to gain an appreciation of how a product, service or environment is experienced by actual users. Designers need to access people's experiences of products, actions and places, using the information as an invaluable source of inspiration. But how can you know how your product or service actually performs if you don't put yourself in the user's shoes? By using and experiencing a prototype, or an existing or new product, you can develop a deeper awareness of the multifaceted experiences that end-users encounter, as well as a greater understanding of a specific problem and/or activity.

This method requires you to leave your studio and try the product/s as your real and/or target users do, and evaluate your proposed design. This might involve going out onto your local streets and purchasing a coffee and muffin in a local shopping centre, withdrawing or depositing cash in a city-centre ATM, or assembling flatpack furniture using only a series of hand tools. Through 'exploring by doing' you can experience the subtle differences between various design solutions, and gain insights that can inspire new design directions and opportunities.

When employing this research method you should ensure that the experience is extensively recorded – keeping a detailed diary of photos and notes. 'Try it yourself' enables you to challenge your preconceptions and provides an opportunity for you to reflect critically upon what you are trying to achieve.

Chapter 3 Learning

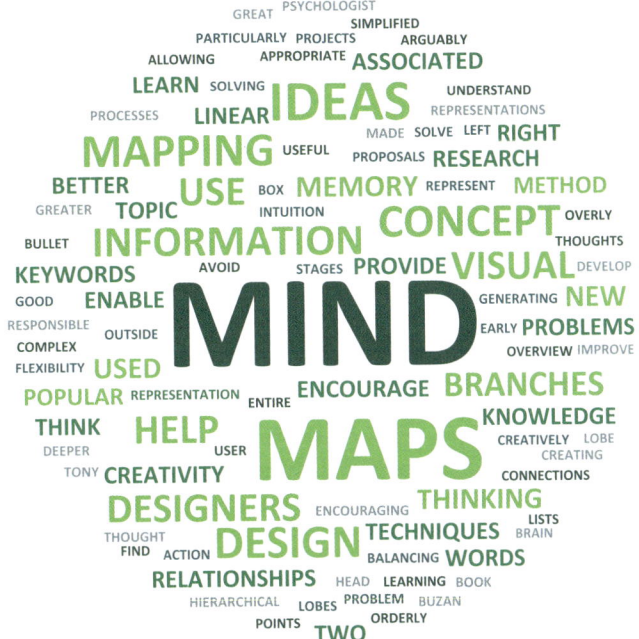

Fig. 9
This 'word cloud' has been generated by the application Wordle™, which gives greater prominence to words that appear more frequently in the source text (here taken from the opposite page); it creates a hierarchy and allows us to visualize the key points.

Fig. 10
A simple mind map showing how words and ideas are generated around a central theme.

MIND MAPPING

A mind map is a visual representation of hierarchical information. Mind mapping, made popular by the psychologist Tony Buzan in his 1974 book *Use Your Head*, is a great method for creating visual representations of words and ideas. It is said to improve memory and develop deeper thought processes by encouraging greater use of the right lobe of the brain (responsible for creativity and intuition), balancing out the use of the left and right lobes. Designers can make good use of mind mapping during the early stages of design projects, particularly when problem solving and generating concept proposals.

Mind maps help you to think and learn better. They help you solve problems creatively and enable you to take action. Mind maps encourage creativity and flexibility, and they help you think outside the box. A mind map can be used to represent an entire concept or an idea with branches of associated thoughts. As with other visual learning techniques, mind maps provide a simplified overview of complex information, allowing a user to understand relationships better and to find new connections.

Mind maps help you to avoid overly linear thinking, which is why they are so useful for designers. They are arguably more appropriate than lists or bullet points, since most design problems are not linear and orderly by nature. They include a central idea or image surrounded by branches of associated topics or ideas. Subtopics are then added to the branches as ideas flow freely. Typically in a mind map, the topic and subtopic text is one to two keywords, to provide a reminder for what the idea is; more information is then detailed in attached notes. Mind mapping is commonly used for brainstorming and note taking. The process of building a mind map is very fluid and non-linear, making the expansion of ideas similar to the natural way of thinking. Symbols and images, along with keywords, are used to retain and recall information quickly. Branches and their associations are often created in different colours to help with identification and memory.

Fig. 11
Sampling is used to gain information from a group of individuals from within a specific population that can then be used to describe the whole population. For example, a sample might be shoppers in a particular area of Japan, such as this precinct in Tokyo.

Another popular diagrammatic research method is concept mapping, developed by Professor Joseph D. Novak at Cornell University in the 1960s. As with mind mapping, this graphically illustrates relationships between information. In a concept map, two or more concepts are linked by words that describe their relationship. Concept maps encourage understanding by helping designers to organize and enhance their knowledge on any topic – designers learn new information by integrating each new idea into their existing body of knowledge.

The use of both mind map and concept map techniques is widespread in the design industry. Imagination and association are the keys to high-level memory and creative thinking and both forms of mapping support this. Since many designers are visual or kinaesthetic learners, these visual approaches enable them to structure, present and evaluate research findings in an effective and successful manner.

SAMPLING

The key aim behind sampling is to gain knowledge and information from a subset of individuals from within a population that can then be used to describe the whole population. A population is a group of individuals, objects or items from which samples are taken for measurement. For example, designers might be interested in a population of car owners or mobile phone users, electronic book readers or university students.

Designers will never be in a position to survey very large populations for two fundamental reasons. First, the costs involved in carrying out very large surveys can be extremely high. Second, populations are dynamic, in that the individuals making up the population change over time. Sampling, therefore, provides several advantages:

1. Costs can be kept low.
2. Data collection can be conducted faster.
3. Since the dataset (i.e. sample) is smaller, it is possible to ensure high-quality and accurate data.

A sample is expected to represent the population from which it comes. However, there is never a guarantee that any sample will accurately denote this population; there is always the possibility that the sample of people that have been interviewed, for example, is unrepresentative of the population. This is usually because of a sampling error. For example, 100 college design students are measured and are all found to be over 1.9 metres tall. It is obvious that this would be a highly unrepresentative sample that would inevitably lead to invalid conclusions. This is an unlikely occurrence because naturally these cases (i.e. college design students over 1.9 metres tall) would be distributed more widely among the population. However, these and other types of sampling errors can occur. There are several methods of sampling to choose from.

Simple random sampling is perhaps the easiest and most ideal, but designers can also utilize the following:

— **Probability methods** e.g. simple random sampling, stratified sampling, systematic sampling and cluster sampling. This is the best overall group of sampling methods to use if you require statistical analyses on the data you collect.

— **Quota methods** e.g. quota sampling, proportionate quota sampling and non-proportionate quota sampling. Quota methods are appropriate when you can determine the number of people you need to sample. For example, when you are studying a number of groups you will need equivalent numbers to enable subsequent equivalent analyses and conclusions.

— **Selective methods** e.g. purposive or behaviour sampling, expert sampling, snowball sampling, modal instance sampling and diversity sampling. Selective methods are best if you are focusing on particular groups of the population. For example, observing how a group of working mothers uses their mobile phones in a specific context.

— **Convenience methods** e.g. snowball sampling, convenience sampling and judgement sampling. These methods are useful when you are unable to access a wider population due to time or cost constraints.

— **Ethnographic methods** e.g. selective sampling, theoretical sampling, convenience sampling and judgement sampling. If you are using one of these methods you will need to use your own judgement to select what seems like an appropriate sample.

Significant savings in design projects can be made if an appropriate sampling method is used, an appropriate sample size is selected and necessary precautions are taken to reduce sampling and measurement errors. Furthermore, the selection and use of an appropriate sample method should deliver valid and reliable information.

TASK ANALYSIS

Task analysis is a user-centred design research method employed to systematically understand the steps and processes that users engage in while interacting with products. This method involves breaking down tasks into their fundamental components to identify workflows, challenges and user interactions in a detailed manner. By analysing how individual users or large teams perform specific tasks in particular contexts, designers gain invaluable insights into user needs, preferences and pain points (troublesome areas), enabling them to create products that are more intuitive, efficient and aligned with real-world use.

A good example of the value of task analysis is in the field of medical device design. By undertaking a highly detailed analysis of how a specific task is performed, designers develop task trainers to train surgeons and clinicians how to undertake complex operations and procedures. This helps to ensure that healthcare professionals can perform procedures reliably and safely.

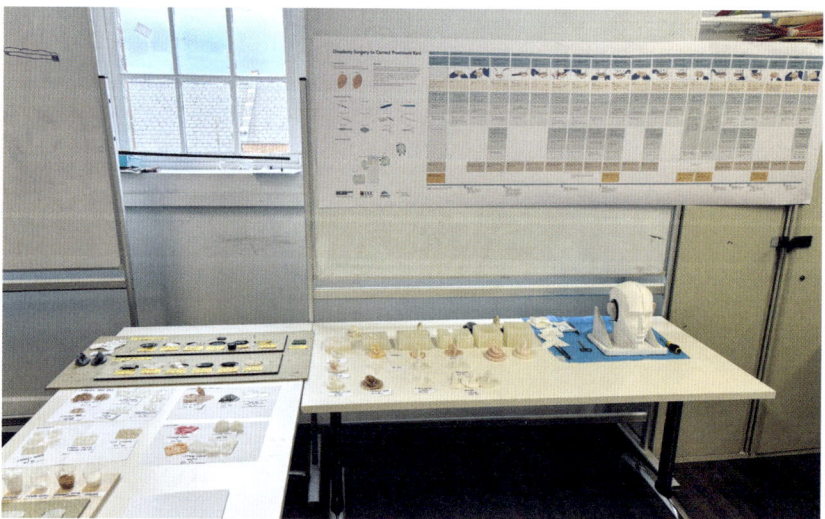

Fig. 12
(Right) Research work developed through a task-analysis process analysing how octoplasty surgery is used to correct prominent ears. The project was undertaken by Leonie Heskin (Professor of Simulation-based Education and Director of ASSERT Simulation Centre); National College of Art and Design, Dublin (NCAD) MSc Medical Device Design students Alexa Broen, Aoife Gallagher, Lochlann O'Regan; and NCAD design staff Enda O'Dowd and Derek Vallence.

Fig. 13
(Opposite) A detail of the Earee Task Analysis, undertaken by Professor Leonie Heskin and NCAD students and staff.

Chapter 3 Learning

Task — **1 | Pre-Incision**

0.0 Subtask Objective: What the subtask aims to accomplish	**1.1 Assess ear** Objective: Orient self to deliver local anaesthetic	**1.2 Add local anaesthetic at lobe, front and back of ear** Objective: Reduce postoperative pain	**1.3 Add two long sutures through Helix** Objective: Allow assistant to hold back ear for access
		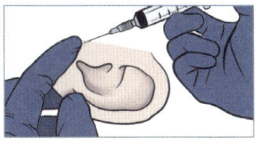	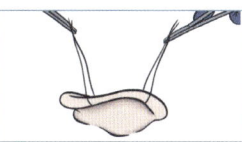
Tools: Tools used and their context (RH=right hand, LH=left hand, RF=right foot) Note: Assume surgeon is right-handed	**Tools:**	**Tools:** Local anaesthetic (RH)	**Tools:** Permanent sutures
P: Perception: What the surgeon experiences in terms of senses. 👁 Visual 👃 Smelling ✋ Tactile 👂 Hearing	**P:** 👁 See patient under sterile drapes with head slightly turned on pillow to expose ear for operation. 👁 See sites around ear to inject local anesthetic.	**P:** 👁 See needle enter ear. ✋ Feel give of skin as needle enters. 👁 See anesthetic create deformation in skin.	**P:** 👁 See needle pierce through skin. ✋ Feel needle go through ear.
C: Cognition: What the surgeon must think about in response to perception.	**C:** Assess ear shape and size. Decide body placement relative to ear for procedure. Identify where to add local anesthetic to ear.	**C:** Avoid inserting needle too far in skin. Decide how many injections to give, avoiding extreme deformation.	**C:** Understand and choose needle placement to give adequate access to incision site.
A: Action: What physical action the surgeon must complete, corresponding with cognition.	**A:** Assess both ears by standing directly in line with patient. Sit on stool on first side of patient to prepare for delivering local anaesthetic.	**A:** Insert needle of local anaesthetic around ear.	**A:** Insert suture needle through ear with black silk thread, joining thread ends with knot halfway.
PCA repeated, as needed.			**P:** 👁 See assistant pull ear back with sutures. ✋ Feel tension of ear pulling back.
			C: Avoid pulling ear too tightly with sutures.
			A: Hand suture threads to assistant to pull back ear.
Risks: Risks that could adversely affect surgery outcome.		**Risks:** Deforming the ear by delivering too much anaesthetic would make it difficult to judge aesthetics.	**Risks:** Placing sutures too close together could impede access to suture site.
Insight: Key considerations for designing an otoplasty task trainer.			
	Time: Timestamp from operation observed on 26/09/24 **Condition:** A notable circumstance in the room	**Time:** Timestamp from operation observed on 26/09/24 **Condition:** A notable circumstance in the room	

CASE STUDY
delaO DESIGN STUDIO'S CARAVANA: DESIGNING A MOBILE COFFEE UNIT FOR SOCIAL IMPACT

Introduction
Caravana is a socially driven brand dedicated to the distribution of high-quality coffee produced by Tzotzil communities from Chiapas, Mexico. The brand's mission extends beyond commerce, aiming to create economic opportunities for indigenous coffee producers while offering premium coffee to urban consumers.

In response to the growing demand for accessible, high-quality coffee in urban environments – particularly in areas where cafés are scarce, such as corporate office parks – Caravana commissioned delaO Design Studio to develop a mobile coffee unit.

Objective
The central objective was to design a visually distinctive mobile coffee unit that was both functional and adaptable to diverse environments, from parking lots outside office buildings to university campuses and private events. The structure had to be light enough for easy towing while maintaining the capacity to function autonomously as a complete coffee shop. The design needed to support the preparation of multiple types of coffee – including americanos, lattes and cappuccinos – as well as teas, tisanes and simple food offerings such as sandwiches.

The unit also had to accommodate such essential equipment as a professional-grade coffee machine, an electric oven, a grinder, a cup dispenser and adequate storage for ingredients and supplies. In addition to functional requirements, it needed to align with the brand's visual identity, designed by an external agency, ensuring consistency in its public presentation.

A key challenge stemmed from the legacy of the brand's previous mobile solution: a modified VW COMBI repurposed as a coffee dispenser. While visually striking, the COMBI presented several operational inefficiencies, including limited space, poor hygiene conditions and negative customer perceptions. The new design aimed to overcome these issues by improving usability, cleanliness and the overall user experience for both baristas and customers.

Methods
delaO Design Studio employed a user-centred design methodology, placing the barista's experience at the core of the development process. Initial field research involved shadowing baristas during daily operations and conducting in-depth interviews to understand their workflows, challenges and needs. These observational studies provided insight into pain points, such as workspace constraints and ergonomic inefficiencies, which informed the design direction.

Informal interviews with customers complemented this research, capturing user perceptions and expectations. This dual perspective, combining the needs of both baristas and consumers, ensured the final design would balance operational efficiency with customer satisfaction.

In the next phase, a co-creation workshop was held at the studio. Full-scale volumetric mock-ups of key equipment (coffee machines, grinders, blenders and storage units) were created, allowing baristas to physically interact with the proposed layouts. This 'play pretend' approach enabled baristas to arrange equipment freely, leading to organic insights on optimal spatial configurations.

The final design was developed from the inside out, prioritizing internal workflows before

Opposite, top: Quick prototyping helped the studio determine the appropriate spatial arrangement and design for the mobile coffee unit.
Opposite, middle: CAD rendering of the mobile coffee unit.
Opposite, bottom: The final design that can be stationed on parking lots outside office buildings and corporations.

addressing exterior aesthetics. Once the internal layout was refined, technical constraints, such as weight distribution, ventilation and power requirements, were integrated into the design. A validation workshop with baristas tested the prototype, ensuring the final configuration met functional expectations.

Special attention was given to the visual identity of the trailer. The distinctive shape and profile of the mobile unit act as a clear differentiator, standing out not only during use but also while in transit through the city. In this way, it functions as a moving brand ambassador, continuously communicating its identity wherever it goes.

Results

The result was a highly adaptable and replicable mobile coffee unit that successfully addressed both functional and aesthetic goals. Its modular design enabled easy scaling, allowing Caravana to expand its reach and increase its social impact by distributing more coffee sourced from indigenous communities.

The trailer's design improved operational efficiency, enhanced hygiene standards and elevated the customer experience. Its distinctive appearance also contributed to brand visibility, making Caravana recognizable in diverse urban contexts.

Ultimately, the project exemplifies how industrial design can serve as a tool for social impact, creating systems that not only solve practical challenges but also foster economic growth for marginalized communities. Caravana stands as a model for how thoughtful design can bridge the gap between local producers and urban consumers while enriching public spaces.

delao.mx/caravana

CASE STUDY
MISCHER'TRAXLER STUDIO'S *ACCESS*: GLOBAL WATER DISTRIBUTION THROUGH GLASSWARE

Introduction
access is a set of six drinking glasses designed by mischer'traxler studio in 2022. Produced by glass craftsmen Vetreria Simone Cenedese and Eugenio Panizzi, the project was commissioned as a limited edition by Punta Conterie for the exhibition *Forms of Drinking*, curated by Elisa Testori. The glasses visually represent disparities in global access to clean drinking water, encouraging reflection on water availability and consumption.

Objective
The project began with the broad topic of water. While other designers were asked to create glassware for cocktails and wines, mischer'traxler studio chose to focus on water as a vital resource. The aim was to create a functional yet thought-provoking glassware series that would translate the importance of drinking water into a tangible form.

Methods
Initially exploring water's visual qualities, the research soon shifted toward its role as an essential element of life. Two design proposals emerged: one on microscopic organisms in drinking water, the other on statistical data related to global water accessibility. The project partners chose the latter to emphasize the unequal distribution of clean water worldwide.

The next phase focused on translating these statistics into glassware using Murano glassmaking techniques. Sketches explored colour options, glass-blowing methods and hand engraving. The final six glasses each represent a specific region, illustrating the percentage of the population with access to safely managed drinking water. The division within each glass translates these figures into a physical form.

A transparent upper section represents the percentage of the population with safe drinking-water access. An amber-coloured base with an engraved surface signifies the portion of people with limited access to clean water, while thin engraved lines on the bottom symbolize the proportion of the population relying on unimproved water sources. Each glass was mouth-blown and engraved, ensuring a unique handcrafted quality that contrasts with the statistical data it embodies. The proportions in each glass visually confront the user with the reality of water distribution worldwide.

The set represents: Sub-Saharan Africa with 30% access to safe drinking water, Europe and North America with 96%, Western Asia and North Africa with 79%, Latin America with 75%, Central and Southern Asia with 62% and the world average of 74%.

Results
access presents an abstract yet tangible way of engaging with global disparities in water access. By integrating statistical information into everyday objects, the project makes an often-overlooked issue more immediate and personal. The varying glass proportions serve as a reminder of how water availability differs across regions.

Beyond its function as a drinking vessel, *access* encourages users to reflect on their relationship with water and its value in a global context. The project highlights the privilege of clean-water access and invites a more mindful approach to its use and consumption.

mischertraxler.com

---

access glassware is a set of six drinking glasses. Each glass represents a specific region of the world, showing the percentage of clean water that is accessible there. The transparent part of the glass shows the percentage of safe drinking water while the coloured part represents limited access to water, and the engraved lines on the bottom stand for the unimproved amount in that region.

Chapter 3 Learning

safely managed access to drinking water

basic/limited access to drinking water

unimproved/no access to drinking water

Sub-Saharan Africa
30%

Europe /
North America
96%

Western Asia/
North Africa
79%

Latin America
75%

Central/
Southern Asia
62%

world average
74%

TUTORIAL

How to write a literature review

Conducting a comprehensive literature review requires you to follow, step by step, a series of important stages:

1. Problem formulation
What is the main problem you are attempting to resolve? Are there any relevant or associated problems? What subjects and/or disciplines have data, information and/or knowledge that might help resolve the main problem and any associated problems and issues? For example, the root of the problem might be an 'energy' issue, and physical scientists and/or engineers may possess relevant information and knowledge. However, you might need to delve deeper to find the information and knowledge you require from more specific scientists and/or engineers, such as geologists, material scientists, mechanical engineers and electronic engineers, for example. You should always have an open mind on the data and information you collect and seek to find competing and opposing data. Are the 'experts' always correct? Is there another way? Why is information from source A so different to that from source B?

2. Data collection
Where will you find the data, information and knowledge that you need to help you undertake the design project? How will you collect it? How wide and deep will you search for it? You need to ensure that you cover as wide a review of the literature as you possibly can in the time available, including published articles, papers, books, reports, academic dissertations, conference papers, trade magazine articles, patent searches, legal reports, analagous product information, statistical data, government and private bodies' reports and market trends' data. All of these would be relevant in a typical product design project.

3. Data evaluation
How will you evaluate the data and information that you collect and review? What criteria will you use for this evaluation? For example, what is the provenance of the information – that is, how authentic is it and what is its quality? Is the report objective? Has the author been unbiased or prejudicial in his or her reporting of the research? How persuasive is the report? What is the value of the data and information that you have found and reviewed?

4. Analysis and interpretation
What are the major findings and/or recommendations of the literature that you have reviewed? What will be the focus of your analysis? Will the analysis be based on quantitative or qualitative data, or both? The analysis and interpretation of the data and information that you have collected during your literature search is an important element of the literature review and should help you structure your report and presentation for the next stage.

5. Literature review presentation
You need to ensure that you cover the literature you have reviewed as comprehensively as possible in your presentation. As a guide, your presentation should include a rich overview of the subject, issue or problem under consideration, along with the objectives of the literature review. You should then divide the work you have found in your review into categories (e.g. those in support of a particular position, those against and those offering different positions entirely), showing how each work is similar to or different from the others.

Literature review

1. Changing times
1.1 Fast fashion
  1.1.1 True cost of fast fashion
1.2 Growth of the value sector
1.3 The high street
1.4 The ethical consumer
1.5 The fashion industry
1.6 Impact of the recession

2. Defining ethical

3. The ethical fashion market
3.1 Market segmentation
3.2 High street availability
3.3 Labelling

4. Consumer purchasing behaviour
4.1 Ethical awareness levels
4.2 Influential factors
4.3 Justification strategies
4.4 Purchasing hierarchy
4.5 The intention–behaviour gap
4.6 The purchasing process

5. The role of the retailer
5.1 Corporate social responsibility (CSR)
5.2 Who controls the industry
5.3 Retailer/consumer communication

6. Conclusions

An example of a structured literature review on the subject of ethical fashion purchasing behaviours (top). The key stages of a literature review (bottom).

TUTORIAL

How to create a great mind map

The beauty of mind maps is that you can either create them using a digital platform, such as Miro, or use the most basic of analogue tools – a sheet of paper and a set of different coloured pens or pencils. When creating a physical mind map, most people find it easier to turn the page on its side and complete their mind map in landscape format – this gives the maximum space for ideas to radiate out from the centre and helps to avoid imposing a subconscious physical hierarchy on the mind map. Whether creating a physical or digital mind map, begin by writing or sketching an image of the main issue in the middle of the page. This can be as simple as a succinct phrase contained within a box or circle.

Next, think up new ideas, action points and statements that relate to the main word(s) or phrase and radiate these out from the central idea. Each new idea should be contained within a box or circle of its own and connected by a line extending in any direction from the main image. You could use a range of different colours so that related ideas and issues are grouped together. The colours you select are not important, but some emphasis should be placed on producing a visually rich and appealing mind map that will interest and stimulate others. You can also enhance your mind map by using different line thicknesses for your branches, arrows, groups, colours and shapes. This process should be repeated as many times as is required to create a number of sub-issues and subtopics.

The completed mind map may be simple, with just two or three subtopics, or it may be a complex mind map comprising tens or hundreds of ideas, sub-ideas, topics, subtopics and connecting lines. Focus on the key ideas, using your own words, and then look for connections between words, ideas and issues. The visual mind-mapping method helps both your creativity and your memory. It will enable you to understand and recall information better, be open to more possibilities and avoid the restrictions inherent in a list format.

Work quickly without pausing to judge what you have written or drawn. Do not edit. Tinkering and working slowly will allow linear thinking and 'analysis paralysis' to set in. Pausing and judging the mind map that you are creating also disrupts the process, which can have negative results. Avoid the notion that things have to be perfect before you can begin.

The key is to think creatively in a non-linear way, working quickly and without worrying about whether everything is correct, appropriate or neat and tidy. In order to create a great mind map, use these simple-to-follow guidelines:

- Write down the central idea or use a central multicoloured image that signifies the mind map subject.
- Think up new ideas related to the central idea.
- Use themes to provide the main divisions of the mind map.
- Enclose each theme with an outline that hugs the shape created by the branches.
- Make sure that the lines that support each key word are the same length as the word and 'organically' connect to the central image.
- Print so that each word used is clear and legible.
- Try to use single key words uncluttered by adjectives or definitions.
- Use colour for vividness and to enhance memory recollection.

If you are creating a digital mind map, it can be really useful to add hyperlinks to the assorted words and images to create another layer of data to stimulate individual and collaborative analysis, synthesis and reflection.

Chapter 3 Learning

71

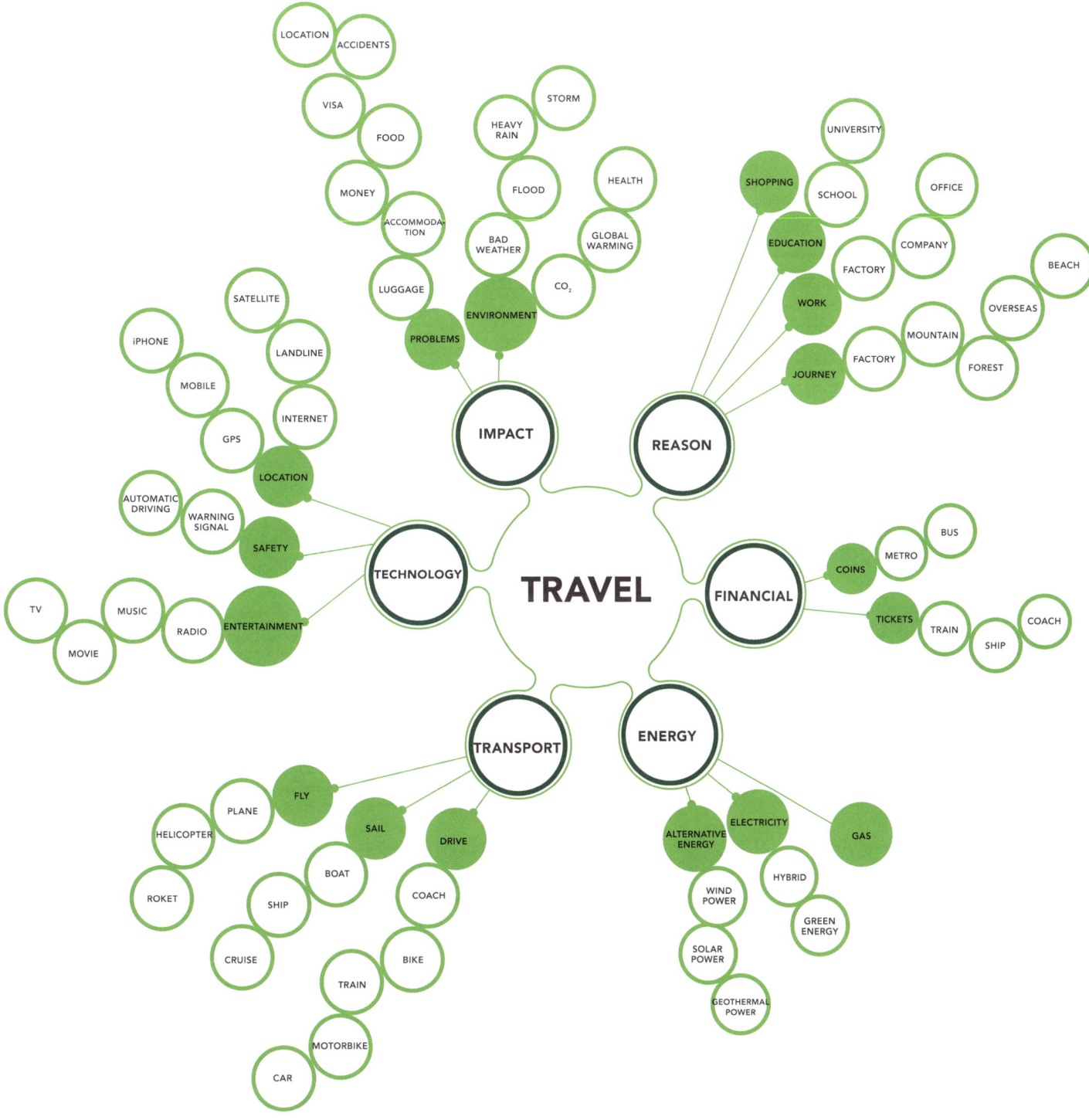

Detailed mind map looking at the motivations and considerations surrounding the central theme of travel.

4 ASKING

Designers can enlist the participation of people in the design process through a range of design research methods that pose questions to help reveal and discover information that can prove invaluable in resolving complex design issues. One of the simplest and quickest ways of eliciting information from individuals is by asking them directly. Using tools and techniques such as questionnaires, focus groups, interviews and the creation of personas, designers can gain a better understanding of the multifaceted relationships that exist between users and the designed products, services and systems they rely on.

QUESTIONNAIRES AND SURVEYS

The words 'questionnaire' and 'survey' are often used interchangeably and confusion can sometimes arise. Questionnaires, however, are basically a printed list of questions, whereas a survey encompasses a range of different elements, including a sample design, a data collection methodology, data collection instruments, analytic techniques, etc. A questionnaire is only one type of data collection instrument. Generally speaking, questions should be clearly stated and there should be a logical flow from one question to the next. To achieve the best response rates, questions should flow from the least sensitive to the most sensitive and from the general to the specific.

Questionnaires and surveys are a relatively simple yet effective way of obtaining information from people. However, one major disadvantage of written questionnaires is the possibility of low response rates. Questionnaires posted out have a response rate of only around one in four returns, although this rate increases when other media formats and approaches are employed. Another disadvantage of questionnaires is the inability to probe responses and, when nearly 90 per cent of all communication between one individual and another is visual, gestures and other visual cues are potentially lost.

This method can be useful, however, for ascertaining particular traits and values of many users relatively quickly. Questionnaires and surveys can be conducted via email, the internet, by post, by telephone and by researchers asking people for responses on the street, in their workplace or at home. Online surveys have had a significant impact on how researchers undertake research. Manufacturers' websites increasingly use devices such as

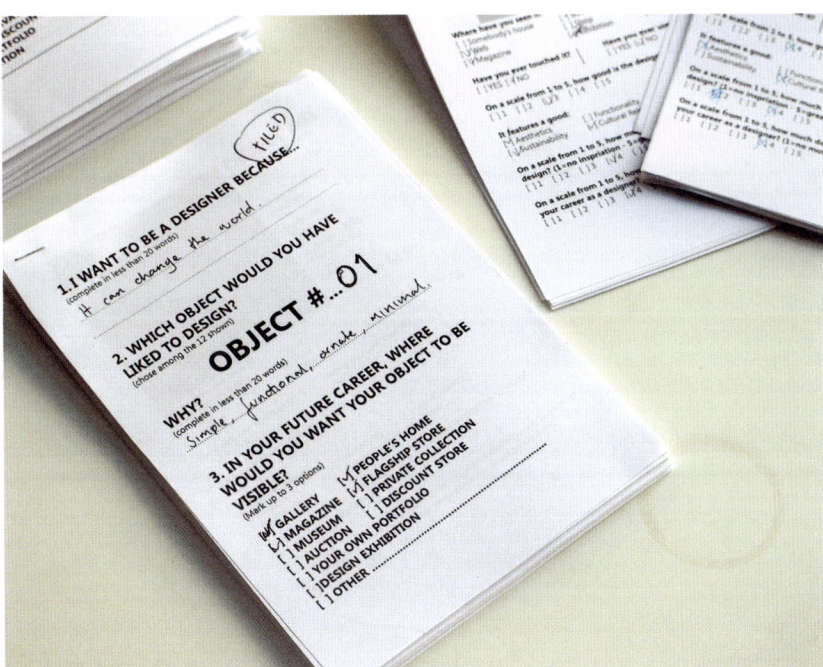

Fig. 1
Example of a printed questionnaire with multiple-option answers to help structure the results.

pop-ups and banner surveys to engage people browsing their sites and gather responses, comments and data from a far broader range of potential customers. There are a number of easy-to-use, customizable templates you can use to construct a questionnaire or survey online, and conduct your own research within minutes.

There are two main types of questionnaire – fixed-response questionnaires and open-ended questionnaires. Fixed-response questionnaires are ones that present a number of alternative responses to a question. Users are asked to mark the option they feel is most appropriate, or rate the options on a sliding scale, perhaps from 1 'strongly agree', through 2 'agree', 3 'not sure', 4 'disagree' to 5 'strongly disagree'.

Open-ended questionnaires ask the respondent to write down their own answers to questions, such as 'What do you like about your bicycle?' or 'Can you suggest things that would improve your bicycle's performance?' Open-ended questionnaires are therefore particularly useful during the early stages of the design process, when the design team might not be entirely sure of the important issues associated with the product being developed.

FOCUS AND UNFOCUS GROUPS

A focus group consists of a number of people brought together in one place to discuss a particular issue or set of issues. The discussions may deal with particular aspects of a designed product, service or system, such as the users' experiences of using a vacuum cleaner, for example. Or they can be more general discussions, exploring, for example, the range of contexts users encounter when using cleaning products in the home. Usually the participants will meet face to face, but focus groups held between people located in different places over large geographical distances using video conferencing are also increasingly common these days.

Focus groups are a form of group interview that capitalizes on the communication between participants to generate information. Although group interviews are often used as a quick and convenient way to collect information from several people simultaneously, focus groups explicitly use

Fig. 2
A focus group on the topic of 'medical care'. Participants are role playing.

group interaction as part of the method. Focus groups typically involve around 8 to 12 individuals who are led by a moderator for anything up to two hours in duration. The group works to a loosely structured agenda, which is an effective way to generate ideas and develop understanding on particular themes without having to reach consensus. The aim is to allow the participants to take the lead in determining the direction of the discussions, while the moderator simply ensures that all the members of the group are able to voice their opinions.

The moderator should also create a set of prompts in case the session loses direction. These prompts should simply be a means of generating further conversation; they should be carefully designed so that they do not lead the participants into giving particular responses. For example, an inappropriate prompt would be 'Don't you find that this handle is difficult to grip?' A more productive prompt would be 'How easy or difficult do you find this handle to grip?' The first prompt is a leading question, while the second is stated in a neutral way.

An unfocus group is a useful method for gaining a number of diverse and sometimes opposing or conflicting interpretations on a given design problem. An unfocus group involves a diverse group of individuals in a workshop-style setting contributing to concept design generation or evaluation activities. Unfocus groups encourage rich, creative and diverse insights which may open up new areas of design activity.

USER NARRATION

User narration, sometimes referred to as 'think aloud protocols', is a valuable method for identifying users' concerns, desires and motivations when using specific products, systems and services. It involves asking users to think about and describe aloud their experience as they perform a specific activity or operate a product in a particular context. Running a good user narration session relies on having something for the participants to interact with, so the organizers should provide an interactive prototype of the designed product, service or system being proposed. Alternatively, the session organizers might wish to utilize existing, competing products that the participants can use and narrate.

Users might be asked to perform specific tasks in a particular order, or asked to explore the product, service or system freely. Giving users set tasks to follow in a predetermined order is useful for uncovering specific issues, whereas user narration in free exploration mode can reveal information on why users use some features of a product and ignore others. During a session the investigator will usually prompt participants to verbalize their thoughts. The prompts might be along the lines of 'What are you thinking of now?' Or they may be more specific and relate to a particular issue or design feature, such as 'Why did you press that button at that time?' Users' narrations can also provide valuable emotional responses to products, which can be prompted by questions such as 'How do you feel when you use the product?'

User narration is an excellent method for understanding not only what problems users have with designed products, services and systems, but also why these problems arise. User narration sessions with small numbers of participants can provide rich data that can lead to better design solutions. One possible disadvantage of this method, however, is the risk of interference between a user's verbalizations, the tasks they are performing with the product and their interaction with the investigator. For example, too much prompting by the investigator can lead to the user making things up in order to be seen to be providing feedback; too little can lead to the user providing less data than might otherwise be the case. Prompting needs to be done in a skilful, careful and balanced way.

Fig. 3
Users 'thinking aloud' while using an interactive product, allowing researchers to understand their concerns and motivations.

INTERVIEWS

One of the simplest ways to explore whether or not users are happy with a product, service or system is to ask them. Interviews basically comprise a series of questions, which are posed directly to the participants. When interviewing users, for example, product designers can ask how they feel when using the product, whether it is easy or difficult to operate, whether they enjoy using it or find it frustrating or annoying. There are three broad categories of interviews that a designer can draw upon – unstructured, semi-structured and structured.

In an unstructured interview the investigator will typically ask each respondent a series of open-ended questions. This gives the respondents opportunities to direct the discussion towards issues that they consider the most important rather than sticking to a predetermined list of questions drawn up by the interviewer. An unstructured interview is an appropriate approach to take, therefore, in situations where the interviewer doesn't know beforehand what the main issues of those being interviewed are likely to be.

In a semi-structured interview, the person asking the questions will usually have a better idea of the main issues surrounding the design project and the questions they wish to ask the respondent. Semi-structured interviews are, by their very nature, a little more constrained than unstructured interviews. The interviewer will usually use prompts to ensure that the specific areas and points they want to cover are addressed, while still allowing for some unstructured contributions from the interviewee.

Fig. 4
Interviews are basically a series of questions posed directly to the participants. Below, architect Rory Hyde interviews designer Ross Lovegrove at the Venice Biennale. On the right, designer Marco Merendi being interviewed at the Cersai fair stand of Rapsel by a H.O.M.E. journalist.

Structured interviews ask respondents to select responses from a predetermined list. For example, an interviewee might be asked to rank features of a new product's aesthetic value on a five-point scale, where 5 means high aesthetic value and 1 means low aesthetic value. This sort of psychometric scale is a device invented by the American psychologist Rensis Likert. Interviewee responses in structured interviews provide data that can be analysed quantitatively, but the issues relating to the product being designed need to be known and well understood beforehand.

Interviews are a flexible method that can be used at various stages of the design process – from helping to formulate specific aspects of the brief and helping to select potential concept ideas through to using prototypes and sharing users' opinions and insights. Interviews eliminate the possibility of misinterpretation between the investigator and the respondent as they usually take place face to face, which is not always the case with questionnaires and surveys, for example.

BE YOUR CUSTOMER/CLIENT

This research method sees the researcher become a consumer in an attempt to actually experience and understand first-hand what real customers feel. It involves leaving the design studio and actually experiencing the entire customer cycle, from evaluating to buying to owning and using the product that you currently design/produce. It enables you to determine which touchpoints deliver a positive experience, and which can be improved.

When undertaking this form of research, designers often take on the role of a mystery shopper. Market research agencies, consumer watchdog organizations and companies use this technique to measure criteria such as quality of service or regulatory compliance, or to gather specific information about their own or competitors' products and services. While gathering information, researchers usually do not declare their activities – they perform their research 'undercover', which can pose ethical issues. Mystery shoppers may sometimes be required to take photographs or measurements, return purchases to evaluate customer service procedures, or record quantitative data such as the number of sales, or the time taken to be served by staff.

Research tools for typical 'be your customer' assessments range from simple questionnaires to comprehensive video recordings. Researchers often complement these field studies by testing the knowledge and service skills of those representing a product, brand or service by asking atypical questions or posing specific challenges to evaluate a range of scenarios. Research data can then be reviewed and analysed to provide quantitative and qualitative statistical analysis reports, helping the design team and company as a whole to design better products, services and customer experiences.

By asking your company or client to describe or enact their typical customer's experience, and then comparing this to the 'be your customer' research findings, you can highlight any reality gaps between a company's internal perception of itself, its offerings and its customers, and those held by its actual customers. You can also undertake stakeholder workshops, where working with existing and/or prospective customers can help identify problems and potential solutions.

When developing a new product or merely refining an existing one, it is vital to challenge your assumptions and place yourself in the shoes of your users. Ask questions, such as 'What value would this new product bring to a prospective buyer?', 'What price would they consider reasonable?' and 'What reasons might they have for not buying?' By becoming your customer, you can fully evaluate a product, from the retail experience of purchasing it, through to use and aftersales support, and thereby generate invaluable detailed data and feedback.

Fig. 5
The design researcher adopting the role of the customer. This method allows the researcher to experience and understand first-hand what customers experience during the entire consumer cycle, from buying to owning and using the product, service or system in question.

BRAND DNA ANALYSIS

The branded products we use and consume say a great deal about our individual tastes and personalities. A brand is an amalgam of product design, logos, slogans, advertising, marketing, packaging and consumer recognition. Designers need to ensure that their products inspire an emotional resonance in consumers, encouraging them to develop a relationship with a brand or product line that evolves over a lifetime of purchasing. As consumers we embrace brands because we feel we are getting more than just a physical product, and we choose from a nuanced family of products that evolve to meet our needs, wants and desires over time. Brands promote consumer choice, and through the development of brand 'extensions', manufacturers are able to target a range of consumers.

Consumers are drawn to branded products because they embody values they are attracted to – such as authenticity or exclusivity – and a range of research techniques have developed over the years to help designers cater for consumers who wish to position themselves socially through the products they purchase.

Brand DNA analysis is a holistic method that aims to reveal the multilayered aspects of a branded product through the integration of data gathered from questionnaires, focus groups and detailed formal design analysis. The technique evaluates a product through a range of perspectives:

Aesthetics
Sensorial perspective. How does the product look, feel, smell and sound?

Interaction
Behavioural perspective. How do users interact with the product? What behaviours does the product encourage?

Performance
Functional perspective. What does a product do? What problems does it solve?

Construction
Physical perspective. How is the product made? What is it made of? What technologies does it employ?

Fig. 6
Selection of leading global brands' logos.

Meaning
Mental perspective. What meanings and emotions does the product provoke?

Brand DNA analysis explores the design language, visual codes and signifiers that characterize a brand, and how a brand's values are translated into a physical design. Products are complex, multilayered creations that communicate to consumers on a number of levels, and brand DNA analysis helps ensure that a product genuinely works within the context of its market competitors, while also conforming to the brand identity of its parent company/manufacturer. Brand DNA analysis can also be used to remodel an existing brand strategy.

Brandscaping
This form of analysis reveals which product is the most representative of a brand, what its most characteristic elements are, and which features are the most important for each design layer. This information can then be used to compare and contrast your product and brand against its rivals through a structured evaluation of a sector – known as brandscaping.

A typical analysis will determine the guiding principles and rules of a brand as seen by consumers from across the target global markets. It will reveal what consumers think about the look, feel and impression generated by a brand's logo, products, retail environments, advertising, marketing and customer service. Such specific cultural and market data is invaluable when determining the future direction of a brand, and drawing up local, regional and/or global brand guidelines – the documents produced by companies to assure the consistent tone and use of brand values and identity.

Brand DNA is an ongoing research process. Brands are living entities, and design researchers need to be continually monitoring evolving consumer attitudes to ensure that a brand's products and proposition respond to these cultural changes while remaining true to the brand's core values.

Fig. 7
Brand development of three logos developed around the name 'Tangible Interactions', created to meet a number of different markets.

MARKET AND RETAIL RESEARCH

Market research is the observation of how rival products are advertised, fitted into the market context and retailed in order to discover how the product field is merchandised overall. Researchers aim to uncover the immediate 'brandscape' of competitive brands surrounding a product and establish what the overall visual impression of the sector is.

Retail research studies how people shop in a particular sector, and how much time they devote to browsing. What elements of the 'design language' in this marketplace seem to be the critical ones used by the consumers in making brand choices? Is this a market balanced between buyers and users – for example, a sector featuring adults buying for children? And does this product sector seem to represent an 'easy buy' for consumers, or do they find it confusing or difficult?

Commonly used market and retail research techniques include:

Product camouflage

This method involves designers modifying a series of existing designs, each with different elements removed. The designers then use a focus group to discuss the saliency of different visual elements. The disappearance of some elements may cause the perception of a product to alter.

Name swapping

Another popular technique, this involves swapping the names and logos on different product designs from the same market, and then discussing if and why the resulting designs are 'wrong' for the branded products being researched.

Fig. 8
Research board by Tom Harper, examining the marketplace and uses for domestic cleaning products.

Drawing from memory

This method reveals the most memorable features of a particular product or sector. Consumers are handed blank paper and are asked to draw particular products from memory. This is followed by an in-depth discussion with the group about what each of them has drawn and why.

Touchpoint analysis

Customers experience a product in many different ways, both directly and indirectly. The means by which customers come into contact with a product, brand or service throughout its lifecycle are called 'touchpoints'. Touchpoint analysis is a research technique that breaks these points down into three distinct areas of investigation:

— pre-purchase (marketing and advertising),
— purchase (retail) and
— post-purchase (product use and after-sales care).

A commonly used tool for evaluating the entire customer journey is the touchpoint wheel, which summarizes all the points of interaction where a customer can be intentionally/unintentionally influenced. The benefits of using touchpoint analysis can include better value for the consumer, greater brand loyalty and an increase in profitability for the retailer.

Fig. 9
The key components of 'brand touchpoint analysis'.

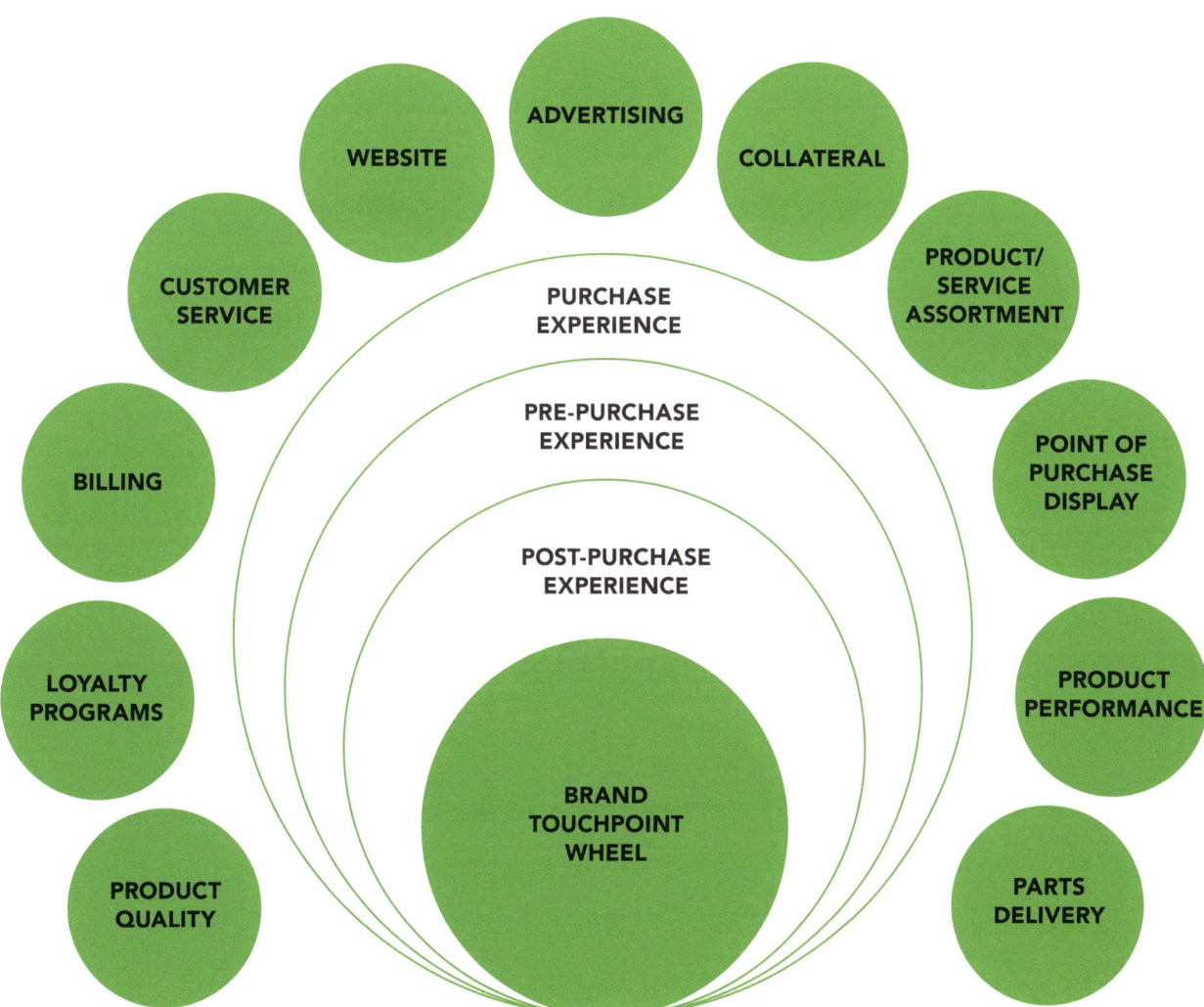

Fig. 10
Inspiration board by designer Rosie MacCurrach for her 'Tales from the Land' collection.

Fig. 11
(Opposite, top) Pantone View Colour Planner contains inspirational photographs and matching colour chips, and forecasts new directions in colour trends.

Fig. 12
(Opposite, middle and bottom) Pages from Andrea Dall'Olio's SS 2010 *Home Interior Trend* book, complete with tactile fabric swatches.

IMAGE AND MOOD BOARDS

Image or mood boards are loosely structured collaged boards that are widely used by designers to portray a range of potential directions for a specific product and/or brand intended for the marketplace. It is important to go beyond the obvious, yet keep the boards credible in the context of the product and brand. Image or mood boards are typically used in conjunction with 2D and 3D drawings and models. They are employed by designers to convey the overall feeling of a project, and involve placing a collection of carefully selected images and objects on a series of physical presentation boards, or canvases on digital collaborative platforms such as Miro, which are intended to inspire, target particular desires, and facilitate creativity and innovation. They may use photographs, illustrations, sample materials and so on to highlight the colour options, typographic possibilities, and the general look and feel of the product, service or system being proposed. Digital mood and image boards also enable the material to be hyperlinked, and provide a way of collaboratively creating, collating and critiquing images, texts and references in real time.

Image and mood boards are especially useful at the early stages of a design project: they can communicate the subjective and emotional aspects of a design to a client and secure their approval before proceeding further. They can also provide designers with feedback before too much time is invested. There is no set formula to creating a successful image or mood board, and they are often intentionally casual. However, carefully taking the client's list of experience characteristics and ideal features into consideration will help to create boards that are appropriate and effective.

Image and mood boards engage stakeholders by facilitating rich discussion and participation between the design team, clients and end-users, particularly during the early stages of the design process. Time dedicated to creating boards at the outset of a project can save time later on in the process.

Chapter 4 Asking

Thus, image and mood boards can be a more cost-effective and efficient way of illustrating a number of design possibilities than other alternative techniques.

PERCEPTUAL MAPPING

When developing a new brand or product line a company needs to select its target market, and then decide how it wants to position its offering(s) within their chosen market segment, or brandscape. This activity is called market positioning and addresses how organizations want their consumers to see their product. Developing a positioning strategy depends on how competitors position themselves. A company needs to decide whether it wants to develop a 'me too' strategy and position itself close to its competitors so consumers can make a direct comparison when they purchase, or whether it wants to position itself away from its competitors.

Perceptual mapping is a research tool commonly used to help develop or evaluate a corporate design positioning strategy, and determine how consumers perceive a brand, product or product range. Through the use of an X–Y axis designers can arrange and plot market research data visually using comparator terms such as cost, quality and impact, and map their target market and audience.

A typical perceptual map might explore consumers' perceptions of a brandscape by evaluating products on the two dimensions of radical/conservative and expensive/affordable. What might emerge from such mapping is that several companies produce products that are perceived by consumers to be positioned close to each other. This would indicate a competitive grouping and crowded marketplace, and a company wishing to introduce a new product might then look for an area of the map free from competitors – a 'gap in the market'. Some perceptual maps provide

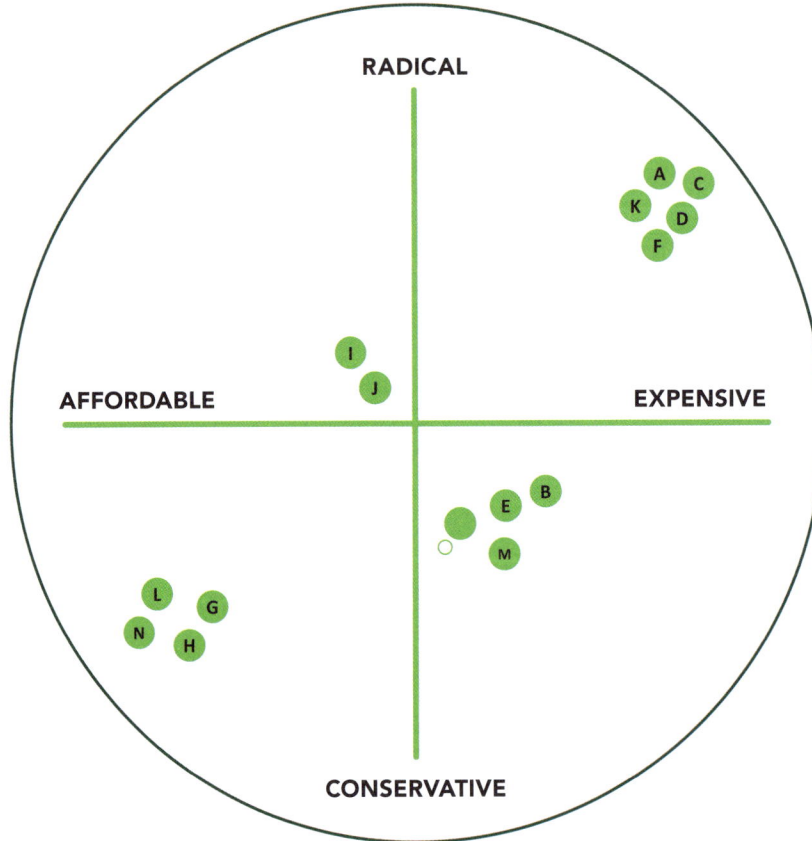

Fig. 13
This perceptual mapping diagram shows how consumers perceive rival products in terms of price (on the X axis) and innovation (on the Y axis). This can help identify overcrowded markets as well as gaps in the market.

additional data by using different-sized dots or circles to indicate the sales volume or market share of the various competing products.

As well as displaying consumers' perceptions of products, a perceptual map can also display the desires of a target consumer group. Consumers can be asked to map their ideal product, through identifying the point where the two dimensions combine to form what is described as an ideal point. By plotting the ideal points of a sample target consumer group, one can identify clusters of high demand, or demand voids. Designers developing new products will usually look for areas of high density of ideal points, and areas without many competitive rivals, in order to locate a fertile market opportunity. Maps are plotted on the basis of consumers' subjective perceptions, so to ensure reliable data, researchers usually collate the views of large focus groups.

Perceptual maps can help identify where a company could launch a new product, by providing qualitative information, such as how consumers perceive their rivals, alongside quantitative information, such as the average price and quality of rival products.

PERSONAS

Personas are fictional characters that are typically based on real-life observations of archetypal users with specific objectives and needs. Personas are created to represent groups of users within a targeted demographic who might all use particular products, brands and services in a similar way. Thus, personas, sometimes referred to as 'character profiles' or 'pen portraits', are a useful method for segmenting different types of users in a crowded marketplace. A persona will typically include specific information such as a fictitious name, age, occupation, educational qualifications, employment history, hobbies, details of their family and friends, and what kinds of products and brands they regularly purchase.

Personas are particularly useful during the exploratory stages of the design process, as they allow product designers to gain a good understanding of their customers' expectations and needs in a relatively cheap and straightforward manner. They help you to understand the people who will be using your product, service or system – and by designing for an archetypal persona you can, in turn, satisfy the broader group of consumers represented by that archetype.

When developing a set of personas you should start with a foundation document that includes references to existing data that informs your initial assumptions. You should then develop the primary personality on this skeleton, fleshing out the character by determining important characteristics such as goals, roles, behaviour, physical attributes, skills, needs, preferences, opinions and environmental context. Taking into account any cultural differences, you should choose values that are meaningful and believable to the target consumer, and ensure the personas you develop are robust enough to be used throughout the product design development process. Remember that a good persona tells a rich and meaningful story, while avoiding clichéd, dated or offensive cultural stereotypes.

Once you have developed your personas, use them to test your product designs in development. A well-constructed persona provides a visual and anecdotal profile based on in-depth research of 'real' users, and they can help you to understand how people might actually use a product. It is important to continually evaluate the cultural and contextual relevance of your qualitative personas with quantitative validation through testing; most researchers 'retire' or 'reimagine' a batch of personas after 12 months to ensure they use on-trend archetypes informed by up-to-date demographic statistics.

Fig. 14
Collage developed in a workshop situation to examine consumer trends.

Despite their ongoing popularity, the use of personas has come in for criticism in recent years. While personas are intended to humanize user data and guide design decisions by representing archetypical users, they can inadvertently reinforce stereotypes and misconceptions if not constructed with nuance and accuracy. If a design persona is based on limited or biased data, it may fail to capture the true diversity of user experiences, potentially marginalizing certain groups or perpetuating harmful assumptions. The process of creating personas can lead to oversimplification, as complex individuals are reduced to a set of generalized traits that may not fully represent their needs, motivations or contexts. This can obscure critical insights necessary for inclusive design.

Ethical issues also arise in terms of consent and data privacy, particularly if real-user data is used without explicit permission. There is a risk that designers may rely too heavily on personas, becoming disconnected from actual user feedback, which can compromise a product's effectiveness and relevance. Therefore, it is essential you engage in ongoing user research, draw upon a diverse set of perspectives when creating personas and remain vigilant against unconscious biases to ensure that your personas serve as effective tools without detracting from the ethical integrity of your design process.

PRODUCT COLLAGE

Product collage involves participants using image and mood boards to support their understanding and perceptions of issues. This image-based method helps participants articulate often complex and interrelated themes, and place them in an explicit context. Collaging is a generative technique, which enables existing and/or potential users to create visually rich data in response to a set of questions or assignments. The users are encouraged to discuss their thoughts and actions during the production of the collage(s), helping to provide deeper insights for the design research team.

A typical collage is a collection of images selected and assembled by the participants in response to a brief set by the researcher. The images provide more than just the visually rich description of a mood board; they also provide the opportunities for analysis and interpretation of the participants' personal narratives, and the socio-cultural interpretation of the images themselves and that of their authors/creators.

Collage enables participants to convey their thoughts and feelings in words, pictures and a combination of the two. It allows participants to articulate their personal narratives, and provides an insight into their feelings, desires and prejudices. It is argued that this technique paints a richer, more emotional picture than text-based methods, such as questionnaires.

Designers conducting research are often frustrated by the constraints of 'traditional' methods of research. For this reason collage is becoming increasingly popular. It allows designers to express their own subjective and intuitive ideas about a concept, and provides a visual outlet for expression and debate. It is an analytic method that can be used to visualize a narrative conceptually; its visual and textual interface can help both designers and consumers to generate visual images to express verbal concepts, and to formulate ways of capturing, structuring and describing the visual experiences that they may have when viewing or using a product. It can help designers 'see' visual research data in a new way, and it provides an opportunity to triangulate these findings with qualitative data from users, and statistical data generated by more traditional research methods.

EXTREME USERS

Extreme users are individuals who are either extremely familiar or completely unfamiliar with a particular product, service or system. Whereas conventional user interviews collect information from users at the centre of an intended target market, extreme user interviews draw upon the perspectives and experiences of these users from the edges or extremities of the target market.

Extreme user interviews are an effective way of highlighting key issues surrounding particular designed products, services or systems – they can often open up unexpected areas and issues, providing insights for future design improvements or opportunities. Extreme user interviews are an extension of the 'lead user' interviews originally developed by Dr Eric von Hippel of Massachusetts Institute of Technology (MIT), and first described in a 1986 issue of the journal *Management Science*.

Generally speaking, the extreme users method involves three major steps:

1. Identification of user needs and trends
2. Identification of extreme users
3. Conduct interviews

Extreme user interviews are based on the notion that innovative products can be designed and developed by identifying leading or extreme trends in the specific marketplace the product is being aimed at. For example, a company seeking to create an innovative product in the area of audio-video equipment may seek out advice and opinions from a variety of extreme types of users, including DJs, musicians, music fans, producers, or others who use, play, listen, write and record music and video as part of their usual day-to-day activities. The network of friends and colleagues of these extreme users could also be a rich source of information.

Lead or extreme users will likely have knowledge and insights that will be 'outside' or 'beyond' the market, and possess more extreme needs than the typical user. By asking, listening and learning from extreme users, designers will create opportunities for coming up with breakthrough ideas and truly innovative products that may not have surfaced using conventional user interview techniques.

JOURNEY MAPPING

Journey mapping is a visual representation that describes step by step how a user interacts with a product or service. The process is mapped from the user's perspective, describing what happens at each stage of the interaction, what touchpoints (both physical and online) are involved, and what obstacles and barriers they may encounter. This method helps designers understand the user experience, identify pain points and discover opportunities for improvement.

Gathering data through qualitative and quantitative research, such as customer interviews, surveys, data analytics and observations, is crucial for learning from and understanding customer behaviours and needs. When mapping the distinct stages of a user journey it is important to identify all the touchpoints, highlighting user emotions and indicating feelings such as frustration, satisfaction or confusion.

By visually representing this research through a journey map you can create an invaluable tool to communicate insights with a design team, and crucially with stakeholders. Validation of the map through feedback from stakeholders and users of the product or service is necessary to ensure its accuracy and relevance. Finally, regular updates to the journey map based on new insights and market trends will help you to keep it aligned with user needs.

The advantages of this method include gaining a comprehensive understanding of the customer experience, which can lead to improved customer satisfaction and loyalty. It fosters collaboration among teams by providing a shared understanding of customer interactions and can identify inefficiencies or gaps in service delivery.

However, there are also disadvantages. Creating a journey map can be time-consuming and resource-intensive, requiring significant research and collaboration. If the map is not regularly updated, it may become outdated and fail to reflect actual customer experiences. As with all interpretive design methods there is a risk of oversimplifying the research findings, and potentially overlooking complex user scenarios or behaviours. Thus it is important to use a range of complementary methods when evaluating and refining user experiences.

USING AND EVALUATING DATA

In an increasingly digital landscape, data has emerged as a fundamental resource for product designers. Its significance lies in the actionable insights it provides regarding user behaviour, preferences and needs. Through data collection and analysis, designers gain a clearer understanding of their target audience, enabling the creation of products that align with user expectations.

Data informs critical decisions across various design elements. By examining user engagement metrics, designers can assess the effectiveness of these elements and identify areas in need of enhancement. Additionally, data aids in recognizing patterns and trends that can guide the development of new features within a digital product. It also facilitates a response to user feedback and satisfaction, pinpointing opportunities for improving the overall user experience.

Various methods exist for collecting data in product design. Surveys, A/B testing, user interviews and analytics tools provide designers with both quantitative and qualitative insights. Once data is gathered, it can help identify user needs and optimize design elements. When determining which data points to prioritize, designers must align their data collection strategies with their design goals, considering the context of their target audience and product specification.

Accuracy and impartiality in data collection are crucial. You can help mitigate biases by framing questions neutrally and using random sampling.

Chapter 4 Asking

As in all aspects of design research, ethical considerations must inform the collection and usage of data, emphasizing transparency and respect for user privacy.

Striking a balance between quantitative and qualitative data yields a comprehensive understanding of user behaviour. A mixed-method approach where you integrate analytics with user feedback enables you to capture both numerical trends and the motivations behind user actions.

Data-driven design is a highly dynamic field, and you should try and stay up to date with best practice and new digital tools and methods.

Fig. 15
Data visualization graphs depicting lifespan against income.

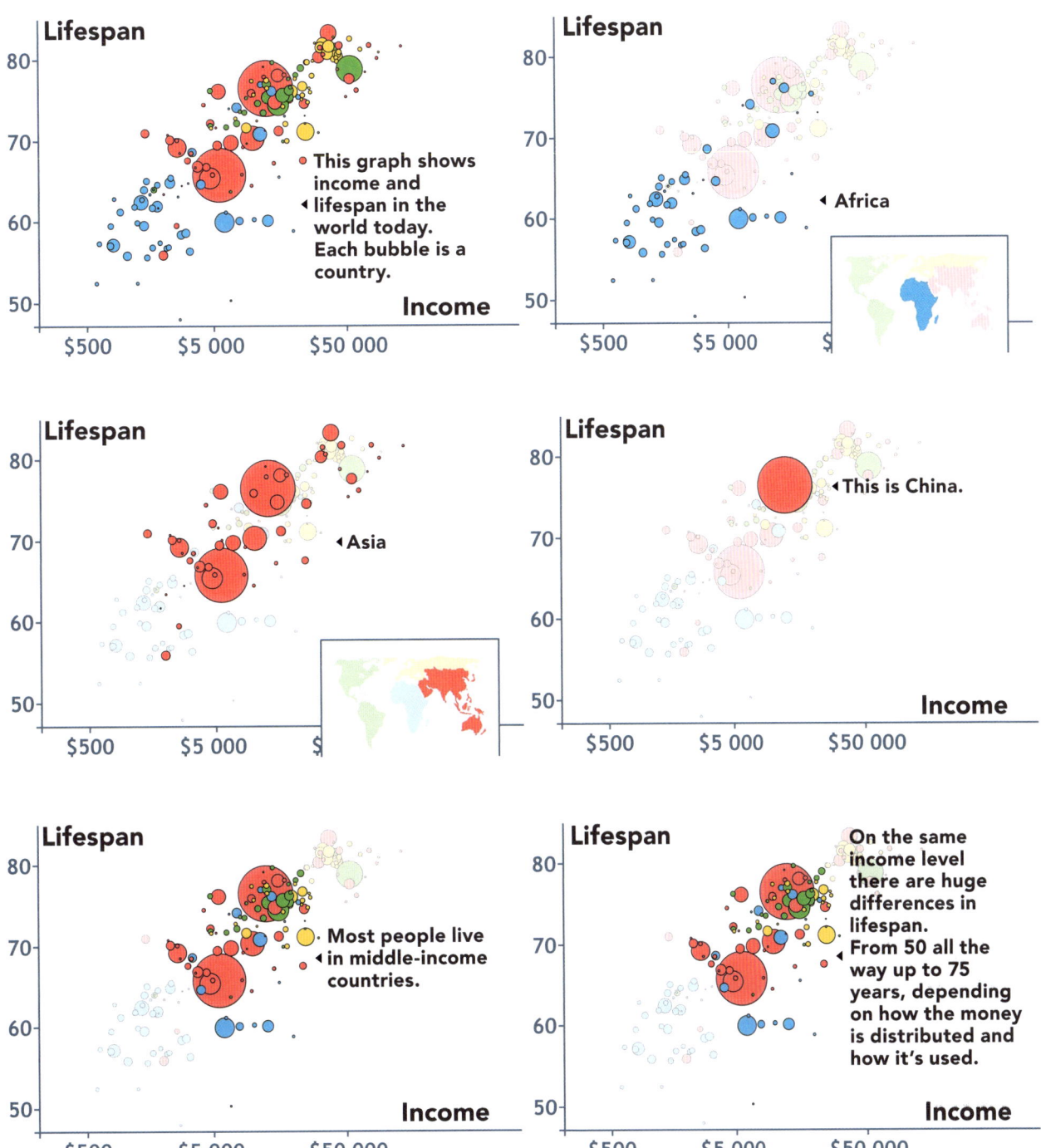

CASE STUDY

KATE STRUDWICK'S FOR.FORM: REDEFINING FORENSIC EVIDENCE HANDLING

Introduction

The current forensic-evidence collection process is fraught with vulnerabilities. When a crime occurs, the first hour – often referred to as the 'Golden Hour' – is critical for preserving forensic integrity. However, current collection methods leave items exposed for extended periods, increasing the risk of contamination or degradation. Compounding this issue, paper and plastic evidence bags limit clear visibility of the evidence, requiring officers and forensic teams to repeatedly open and close them to examine or present the evidence, further jeopardizing its reliability.

Additionally, inconsistencies in evidence handling and manual documentation create delays and open up opportunities for tampering. With budgets often limiting the number of exhibits that can be submitted in court, each item must be meticulously preserved to hold up under scrutiny. With DNA technology becoming increasingly sensitive, minimizing contamination at every stage of evidence handling is more crucial than ever.

Objective

For.Form wanted to design a packaging system that directly addressed these issues by reducing exposure, securing evidence with a transparent yet form-fitting material and integrating a tamper-proof seal. The system would prevent unnecessary handling and movement of the exhibits within the packaging to ensure that critical trace evidence, such as fingerprints and microscopic fibres, remains intact throughout the chain of custody.

For.Form also explores ways to modernize forensic tracking, replacing the manual record-keeping process with RFID (radio-frequency identification) technology to streamline evidence management, reduce delays and enhance security.

The For.Form system is designed to address three major pain points in forensic evidence handling:

1. **Protection**: reducing the window of exposure at the crime scene that allows contamination.
2. **Packaging**: ensuring evidence is contained securely without compromising trace materials.
3. **Tracking**: creating a tamper-proof chain of custody using embedded RFID technology.

By tackling these areas holistically, the goal was to create a seamless solution that detectives, evidence handlers and forensic analysts could trust.

Methods

The project was deeply rooted in ethnographic research within police stations, focusing on evidence handling and forensic workflows, working alongside detectives, crime-scene managers and forensic scientists to identify key problem areas. The research methodology included:

- **On-site observations**: visiting evidence lockers at a police station and courtrooms to analyse the current exhibit-handling processes.
- **Collaborations with key stakeholders**: conducting interviews and co-design workshops with detectives and crime-scene managers to ensure usability and procedural compliance.
- **Material science and prototyping**: working against a strict list of requirements developed through research with forensic scientists and biologists, For.Form experimented with novel polymers and flexible packaging

Chapter 4 Asking

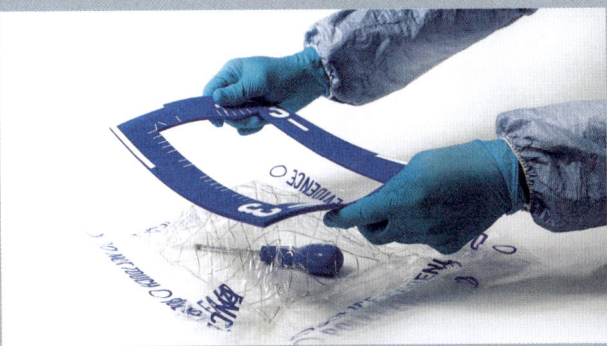

solutions that could be moulded and sealed without compromising biological evidence.
- **Simulated use-case testing**: assessing how officers interact with packaging in a controlled environment and refining the system to match real-world workflows.

Results
For.Form made a number of key breakthroughs:

- **A flexible, mouldable material**: a composite material can be shaped around evidence without distorting fingerprints, DNA or trace residues.
- **Dual-purpose protective dome**: at the point of collection, the dome acts as an instant shield against environmental contamination. Later, it becomes the evidence's permanent packaging.
- **RFID-enabled tracking**: unlike handwritten evidence tags that are prone to misinterpretation and loss, For.Form's embedded chip automatically logs every interaction, guaranteeing forensic integrity.
- **Tamper-proof sealing mechanism**: a closure ensures the package cannot be opened without authorized forensic tools.

katestrudwick.com

For.Form brings design thinking and material innovation to police detectives to redefine the process of packaging and processing forensic evidence.

CASE STUDY

PARSONS & CHARLESWORTH'S CATALOG FOR THE POST-HUMAN

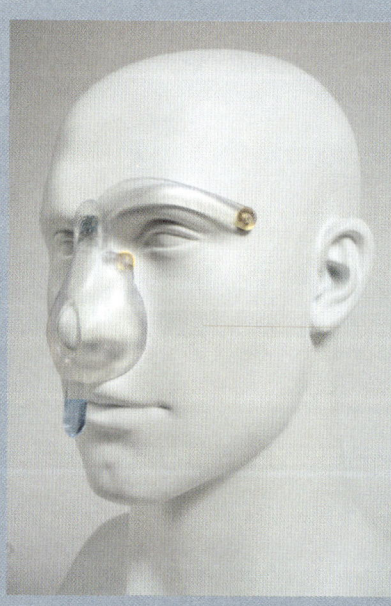

Below: RemWake, a smelling-salts alarm clock.
Bottom: MycoPops, a range of probiotic lollipops.

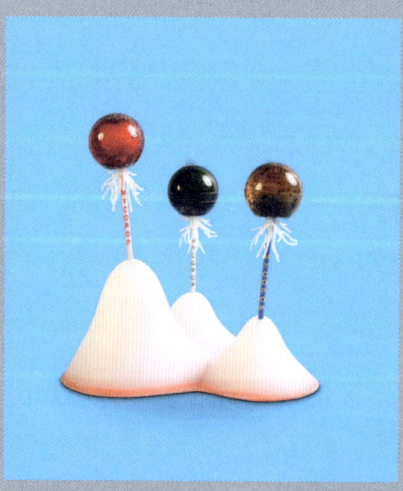

Introduction

Catalog for the Post-Human is a satirical multimedia installation by research-based studio Parsons & Charlesworth. It was presented at the 2021 Venice Architecture Biennale as a collection of sculptural works and animations with the appearance of a near-future tech organization's trade-fair booth. This immersive work features ten quasi products – objects that look like real consumer products but were in fact fictional prototypes – designed to provoke conversations about the impact of enhancement and surveillance technologies upon an increasingly contingent workforce.

Objective

The trade-fair installation aimed to engage viewers with issues relating to the ethical and social dimensions of our technologically mediated futures, especially the role of tech and wellness companies in promoting an 'always on', productivity-centric lifestyle.

Methods

Although the work was for a cultural venue, Parsons & Charlesworth adopted methods from corporate design consulting. They worked through such phases as a literature review, expert interviews, mapping, persona generation and image boards to generate material to inform each fictional product.

Interviews with experts in workplace management, cognitive-enhancing drugs and artificial intelligence led to insights that were mapped alongside emerging societal and workplace trends. These described what has been called 'the rise of the gig economy': a decrease in conventional salaried jobs, fewer benefits, restrictive contracts, an older retirement age, increased workplace surveillance, etc. This research mapping enabled Parsons & Charlesworth to identify six themes that appeared most prescient:

1. Productivity
2. Training
3. Health/Medication
4. AI Collaboration
5. Surveillance/Monitoring
6. Profile/Reputation

They then created eight personas, each with an imagined job, accompanying skillset and the pressures or stresses it entailed. The eight personas were:

- The Best Selfer
- The Credit Junkie
- The Cognitive Enhancer
- The Microbiome Obsessive
- The Bio-hacker
- The AI Collaborator
- The Identity Switcher
- The Extreme Empath

Using these personas as target customers for the fictional client Catalog for the Post-Human, they imagined the kinds of enhancement-technology products each persona would need to stay competitive in an increasingly automated workplace.

For example, the Cognitive Enhancer, described as 'recognizing the variety of cognitive states required for each job and the drugs needed to get them into – and out of – that state', led to a section of the final exhibition called Cognitive Management, which incorporated two products: the NootDial™, a dispenser for nootropic pharmaceuticals, and the Morning Ritual™, a kit for microdosing LSD. The Microbiome Obsessive, defined as 'following extreme health regimes to maximize productivity and as insurance against illness and loss of income', led to the creation of an Optimized Wellness

NootDial, a dispenser for nootropic pharmaceuticals.

area for tools to help you maintain hydration and a healthy gut, which included MycoPops™, a range of probiotic lollipops containing a type of bacteria found in soil.

Results
The installation provided the audience with a glimpse into a dystopian near-future in which we may be forced to enhance ourselves. The application of research methods typically used in industry gave the fictional products a clear context and, despite their satirical nature, some visitors to the exhibition believed them to be real products.
cftph.work

TUTORIAL

How to create a great questionnaire

It's important to give respondents enough time and an adequate place to complete their questionnaire.

Before setting off on writing your questionnaire you should use the following steps as a guide:

1. Decide what information you want to gather from the questionnaire.

2. Keep the questionnaire as short as possible, asking only those questions that will provide the information you need.

3. Use a casual, conversational style, making the questions easy for almost anyone to understand.

4. Structure the questionnaire so that the questions follow a logical order and evolve from general to specific.

5. Use multiple-choice questions whenever possible. This helps the respondent to better understand the purpose of your question and will reduce the time it takes to complete the questionnaire.

6. Avoid leading questions that might generate false positive responses. For example, the question 'How great was the service provided by our excellent waiters?' should be 'How was the service provided by our waiters?'

7. Use the same rating scale throughout. For example, if the scale is from 1 to 5, with 5 being the most positive, keep that same scale for all of the questions requiring a rating.

8. Test the questionnaire on 10 to 15 people before you produce it for mass distribution. Conduct an interview with each of those respondents after he or she completes the survey to determine if your questions were easily understood and easy to answer.

Following these simple steps will help you get the most out of your questionnaire and go some way to obtaining the information you require from people.

QUESTION SEQUENCE

In general, questions should flow logically from one to the next. To achieve the best response rates, questions should flow from the least sensitive to the most sensitive, from the factual and behavioural to the attitudinal, and from the more general to the more specific.

Basic rules for questionnaire item construction
— Use statements that are interpreted in the same way by members of different sub-populations of the population of interest.
— Use statements where people that have different opinions or traits will give different answers.
— Think of having an 'open' answer category after a list of possible answers.
— Use only one aspect of the construct you are interested in per item.
— Use positive statements and avoid negatives or double negatives.
— Do not make assumptions about the respondent.
— Use clear and comprehensible wording, easily understandable for all educational levels.
— Use correct spelling, grammar and punctuation.
— Avoid more than one question per item (e.g. 'Do you like strawberries and potatoes?').
— Write down the central idea or use a central multicoloured image that signifies the mind map subject.
— Think up new ideas related to the central idea.
— Use themes to provide the main divisions of the mind map.
— Enclose each theme with an outline that hugs the shape created by the branches.
— Make sure that the lines that support each key word are the same length as the word and 'organically' connect to the central image.
— Print so that each word used is clear and legible.
— Try to use single key words uncluttered by adjectives or definitions.
— Use colour for vividness and to enhance memory recollection.

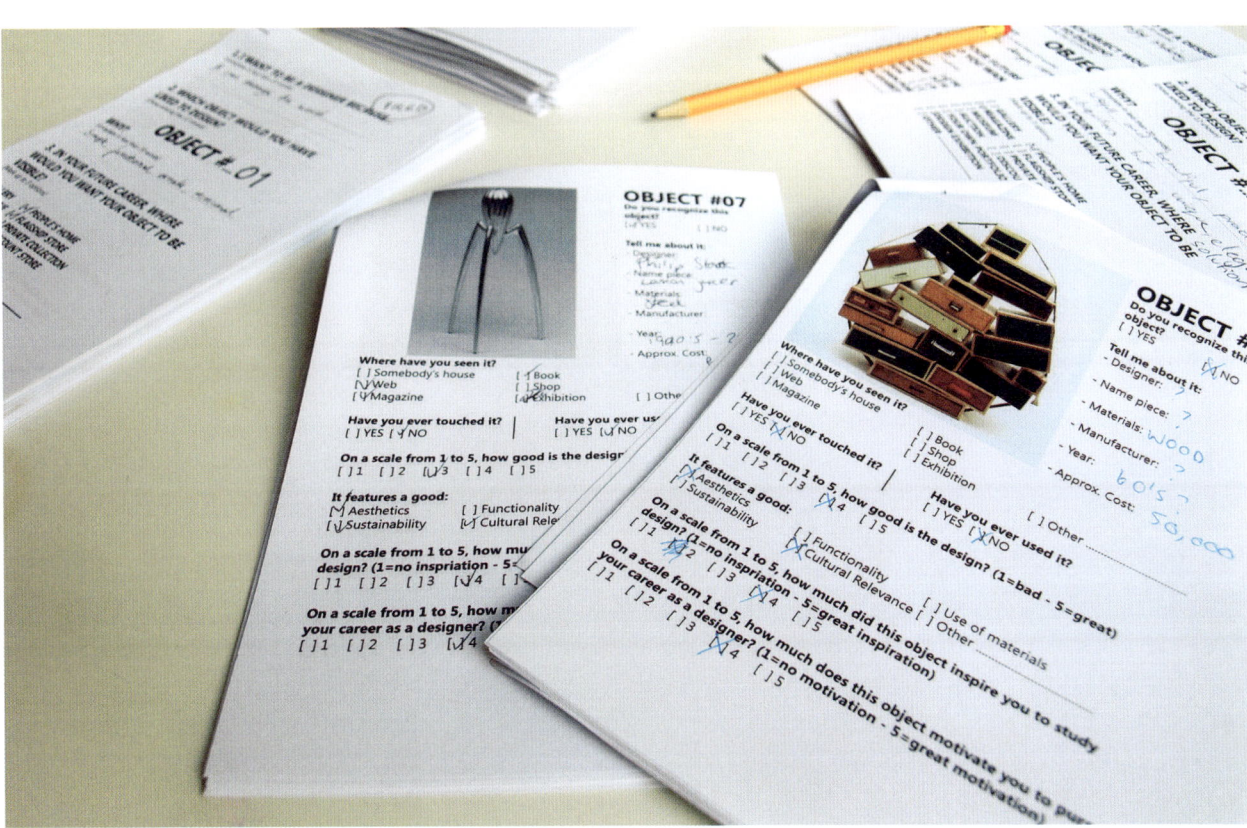

Completed 'I want to be a designer because...' questionnaires.

TUTORIAL

How to conduct great interviews

Interviews are a commonly used research technique, enabling design researchers to pursue in-depth information around a topic of research, and/or to follow up on the results of a questionnaire. Interviews can follow a number of formats:

Informal, conversational interview
This format doesn't use a set of predetermined questions, and adopts a flexible 'go with the flow' approach.

General interview guide approach
This format ensures that the same general areas of information are collected from each interviewee; this provides more focus than the informal approach, but still allows a degree of freedom and adaptability in getting information from subjects.

Standardized, open-ended interview
In this format the same open-ended questions are asked of all interviewees.

Closed, fixed-response interview
This highly structured format asks all interviewees the same set of questions and asks them to choose answers from among the same set of alternatives.

Video still images of 30-second face-to-face interviews conducted by Johnny Weir.

Step-by-step guidelines

— Choose a setting with no distractions and where the interviewee will feel comfortable.

— Explain the purpose of the interview.

— Address terms of confidentiality, explaining who will get access to the respondent's answers and how their answers will be analysed. If their comments are to be used as quotes, you must get their written permission to do so.

— Explain the format of the interview you are conducting and its nature. Indicate how long the interview usually takes.

— Tell the interviewee how to get in touch with you later if they want to.

— Ask them if they have any questions before you start with the interview.

— Ask for permission to record the interview, as you can't count on your memory to recall their answers.

— Get the respondent involved in the interview as soon as possible.

— Before asking about controversial matters (such as feelings and conclusions), first ask about facts.

— Intersperse fact-based questions throughout the interview – this will avoid long lists of fact-based questions, which tend to leave respondents disengaged.

— Ask questions about the present before questions about the past or future.

— Occasionally verify the tape or video recorder (if used) is working.

— Ask one question at a time.

— Attempt to remain as neutral as possible and don't show strong emotional reactions to the interviewee's responses.

— Encourage responses with occasional affirmations and nods of the head.

— If you take notes, be careful that this is not revealing or distracting.

— Provide a transition between major topics, for example 'We've been talking about X and now I'd like to move on to Y.'

— Don't lose control of the interview and allow the respondent to stray to another topic, take too long to answer or even begin questioning you.

— You should finish the interview by allowing the respondent to provide any other information they wish to add and their overall impressions of the interview.

— Make any additions to your written notes to clarify your findings as soon as possible after the interview to ensure you don't forget any crucial points.

— Write down any observations made during the interview that may have impacted on the process.

5 MAKING

Chapter 5 Making

Designers make models and prototypes to inform their design and decision-making processes. While these have traditionally been perceived as highly developed physical models, contemporary designers now use both terms to describe any kind of representation that is created to help designers, users and clients to understand, explore and communicate the qualities of a product, and how a user might engage with it. Thus, the terms 'models' and 'prototypes' are nowadays used to describe a range of design representations, from concept sketches through to a variety of physical, CAD and virtual models that explore and communicate design propositions and contexts.

Research through making underpins design practice, and the act of experimenting through prototyping makes research tangible. Increasingly designers are also considering how they can unmake. Responding to the unsustainable nature of much product design, designers are researching strategies such as upcycling, repair and hacking to avoid perpetuating an extractive or unsustainable model and move towards a circular or regenerative one.

SKETCH MODELLING

Product design is a three-dimensional discipline, and while the immediacy of marker renderings and the visual gloss and ease of CAD offer huge possibilities, it is essential that designers model their concepts physically and test them in the real world.

Sketch models are full-size or scale models that aim to capture the embryonic ideas emerging from the design team's initial concept development. These expressive and rapidly produced models will progress in complexity, resolution and finish until the designer or team are confident enough to progress to more time-intensive models. Sketch models are typically hand-carved or sculpted from readily available materials such as urethane foam or foam board. Due to the materials used, they are generally crude and not fully representative of the final intended design.

Sketch models enable designers to visualize their two-dimensional designs three-dimensionally. They provide an insight into sculptural aspects of a product's evolving form, and allow for quick and effective evaluation of aesthetics, ergonomics, functionality, usability, proportion, and packaging and configuration options. Designers can then develop these aspects further, as required.

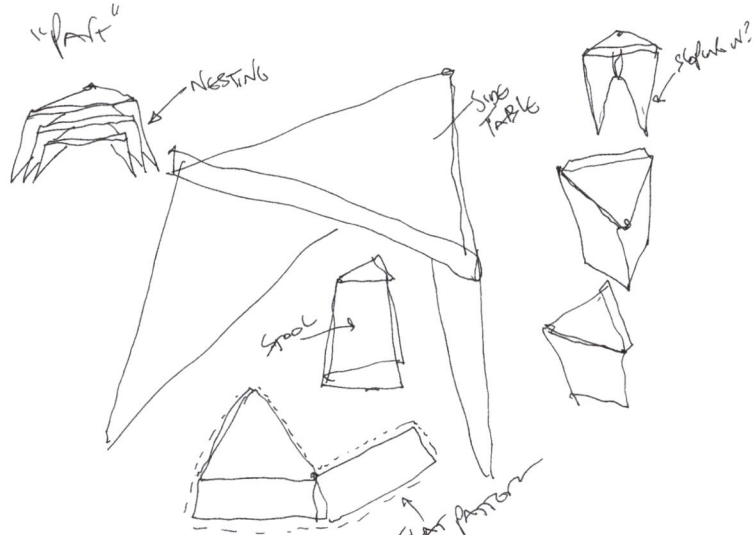

Fig. 1
Early concept sketches are quickly worked up into full-scale cardboard models by Stephen Burks during the design process for his Handmade Furniture tables. Burks believes working with full-scale models is crucial to understanding scale and usability.

Sketch models also help designers to convey their designs to others in a design team or as a final representation of a design to a client. They can be used to test public reaction to a new design, and evaluate its suitability within a market. They can also be used to test the structural integrity of a design, or to test a particular part of a design, such as a mechanism.

Sketch models are almost always produced to scale. This can be either smaller than actual size (i.e. 1:5, 1:10, 1:20, 1:50 or 1:100) for large items such as pieces of furniture or interiors, actual size (i.e. 1:1) or larger than actual size (i.e. 2:1 or 5:1) for very small products or for developing new mechanisms. The scale of a model also depends on the stage of development. In the early stages of a design project, when many ideas are being explored, smaller-scale models are more common.

MOCK-UPS

A mock-up is a life-size physical model constructed from easily fabricated materials such as rigid card, wood and foam. Such models are used to evaluate the physical interaction, scale and proportion of product design concepts. Mock-ups and simulations are most commonly used to evaluate designs during the early stages and the midway point of the design process.

Mock-ups are commonly required for bespoke furniture products, enabling designers to produce a full-size replica using inexpensive materials in order to verify a design's form, scale and ergonomics. They are often used to determine the proportions of a design and how it relates to a spatial context. Mock-ups can also be used to test the colour, finish and other specific details – factors that cannot be easily visualized or resolved through sketches or technical drawings.

Mock-ups that replicate a mechanical action or enable a physical property of a design, such as its strength, stiffness, comfort or durability, to be tested are known as test rigs. While computer modelling and analysis techniques provide designers with important insights into how a product component might perform, they are based on assumptions and approximations of actual product behaviour.

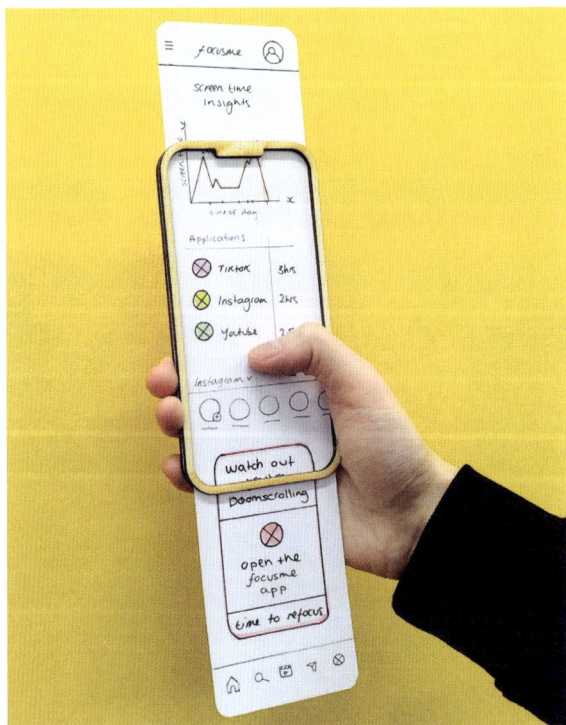

Fig. 2
Simple 2D paper mock-up, by Nuria Mora Hurtado, created at actual size, for a new mobile phone application concept.

Chapter 5 Making

Fig. 3
(Above) Cardboard and polystyrene models of Edward Barber and Jay Osgerby's Pavilion Chair. These were created at an early stage of the development to test comfort and stability.

Fig. 4
(Left) CAD prototype of a digital scrapbook. Digital prototypes (CAD models) are a relatively quick and effective way of testing concepts without having to physically make them.

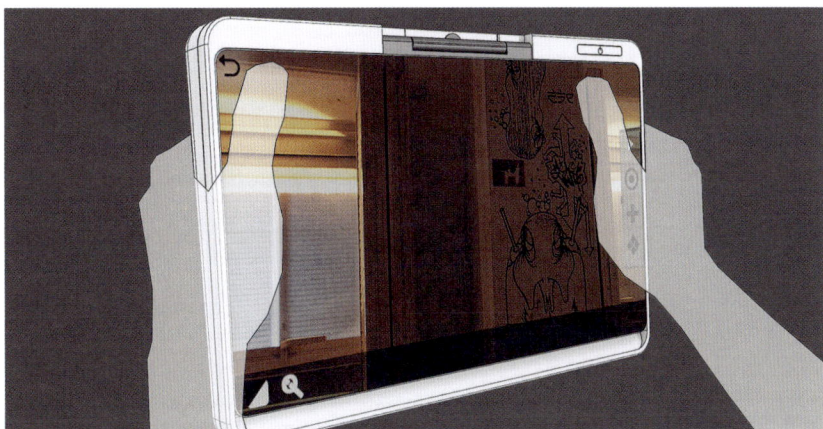

Full validation of a design can only be achieved by the extensive practical testing of representative test rigs and mock-up prototypes. Such testing is usually undertaken by engineers working closely with the design team to ensure that the learning gained through the tests is fed directly into the integrated development of a product for production.

Mock-ups that do not incorporate any product styling and are only intended to demonstrate the basic mechanism of a product are typically referred to as proof of concept models. These are used to 'prove' the viability of a potential design approach, such as a product's range of motion, mechanics, sensors or architecture.

Mock-ups are a key development and standards testing tool, helping designers to validate design options and determine where further development and testing are necessary. They also provide the engineering team with the opportunity to do a final check for design flaws or possible last-minute improvements. The cost of producing mock-ups is often outweighed by the savings made by avoiding going into production with a design that needs subsequent improvement.

The number of mock-ups required varies from project to project, and depends on the scale and, indeed, the budget available. However, the need to evaluate a product's form, composition, materiality and production processes can only be properly met through intensive prototyping. The development of the Dyson DC01 vacuum cleaner (p.110), for example, required thousands of prototypes before innumerable issues were resolved and the concept reached production.

PAPER PROTOTYPING

Paper prototypes provide a quick way to visualize, organize and articulate basic design concepts. They are widely used in the design and development of new products. Designers can use this method to sketch out and evaluate the basic functionality and usability aspects of their concepts, and assess whether or not user interfaces meet users' expectations and needs.

Paper prototypes are meant to have a 'throwaway' quality. They are intended to be rough, hand-drawn, sketchy representations of a product's interface. Some paper prototypes can be hand-drawn, while others use printed screenshots. Although aspects of paper prototyping can appear rather simple – crude, even – this method of usability testing can result in significant insights and valuable feedback.

Paper prototyping is not a new research method. It has its origins in the mid-1980s, when companies including IBM, Microsoft and Apple started adopting the technique in the design and development of many of their products. Nowadays, paper prototyping and digital paper prototyping are widespread in the evaluation of products with a significant degree of screen-based user interaction. Paper prototypes are also often used in conjunction with eye-tracking software, which can record a user's eye movements as they perform key tasks while using a product, to evaluate and refine new computer interfaces.

Paper prototyping can be used for virtually any type of human–computer product interface, such as a web-based service or a mobile hand-held device. The main aim is to get rapid feedback from intended users while a design is in its early stages of development. Usability testing involves recruiting and selecting a suitable number of users who represent the target market(s), and having them perform realistic tasks using the prototype. A session typically includes a facilitator (someone trained in usability) to run the session, with members of the design and development team as observers, taking notes about what works well for the users and what confuses them.

In addition to usability testing, product design and development teams also find that paper protyping can be a useful way of generating design ideas and conducting internal interface reviews.

Paper prototyping has a number of advantages over other forms of prototyping:

— It is a very fast way to mock up an interface.
— It can detect a wide variety of problems in an interface, including many serious issues.
— It allows an interface to be refined based on user feedback before any real physical implementation begins.
— It facilitates a multidisciplinary approach.
— It encourages creativity from the design and development team as well as the end-users.

Fig. 5
Development models for a new range of ceramic ware, including concept sketches, paper prototypes and clay models.

QUICK-AND-DIRTY PROTOTYPES

Quick-and-dirty prototypes are used as a quick way to communicate a concept design idea to other members of the product design team. From this the team can then evaluate, reflect and refine its ideas before progressing further. The prototypes are built quickly and with any materials that may be to hand – the focus here is on speed rather than quality. Many design projects have short timescales with very demanding deadlines; quick-and-dirty prototypes can help designers 'cut corners' in order to save both time and other valuable resources.

The 'rough-and-ready' nature of quick-and-dirty prototypes encourages exploration too – design teams will more readily add, remove or change elements of a design that they have invested less time in and that has less emotional significance than a more 'polished' prototype. The process is cheap, quick and open to interpretation. It enables designers to focus on the essence of a concept, and avoid getting bogged down trying to resolve details.

Quick-and-dirty prototypes provide 'good enough' or approximate results rapidly. Generally, these prototypes will be used in conjunction with established evaluation methods such as focus groups, interviews or field observations, but applied in a less formal way than is usual. For instance, to save time the design team may cut some corners in the selection and recruitment of participants, reduce the number of users providing feedback and/or reduce the scope of the product proposal's evaluation.

Product concepts evolve through a number of iterative design and evaluation cycles. That is, designers create prototypes and evaluate their strengths and weaknesses, often by assessing verbal or written descriptions of the proposed product design against users' requirements to evaluate their suitability or otherwise. Quick-and-dirty prototypes are an enormously valuable tool for designers during this process. Generally speaking, they will be constructed using a range of inexpensive materials such as paper, card, glue, sticky tape, wood and polystyrene foam, although it is increasingly common these days for many product proposals to be simulated via screen-based interactive prototypes using off-the-shelf computer software, with companies offering specialist product design services in computer visualization and modelling.

Fig. 6
The creation of quick-and-dirty prototypes is an enormously valuable tool for helping designers during the iterative design and evaluation cycles. Shown are various quick-and-dirty prototypes for a new camera concept by Gregor Whyte.

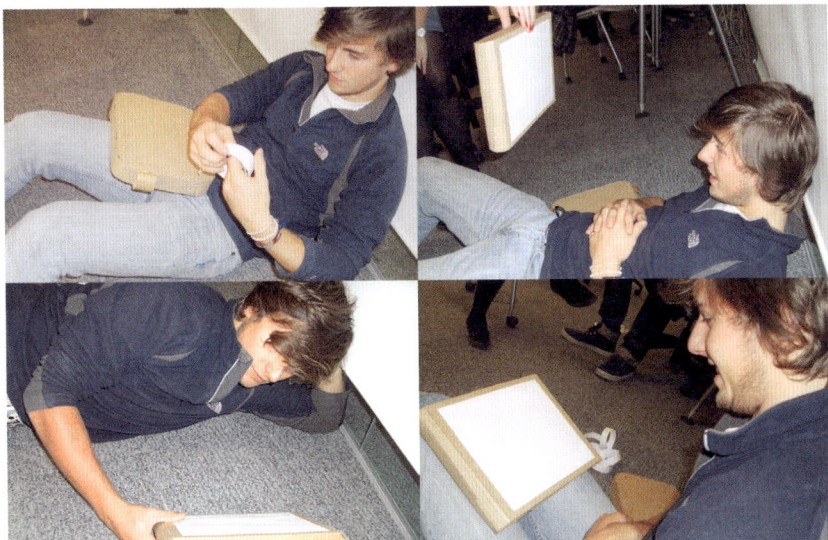

Fig. 7
Quick-and-dirty development prototypes exploring the concept of computing on the go.

EXPERIENCE PROTOTYPING

A recent development has been the widespread use of prototypes that move beyond aesthetics and instead focus on the experiential aspects of a design. Experience prototyping, as it is known, is a useful research tool for detecting unanticipated problems or opportunities as well as for evaluating ideas.

A new generation of smart products, and the integration of digital technology into traditionally analogue products, has resulted in designers needing to prototype physical and digital touchpoints to test and evaluate ideas and assumptions. Digital products and interfaces can be prototyped using wireframes, mock-ups or clickable prototypes. The key is to build prototypes that are easy and fast to make, and that allow designers to get feedback from their customers or users as early as possible. This allows the designer to validate ideas and learn from mistakes through an iterative process to ensure an appropriate user experience that spans the physical and digital aspects of a product.

The discipline of product design has been transformed by the move from a manufacturing economy to that of an experience economy, with experience itself becoming the product. Consumers no longer merely consume products but lifestyles, with products not simply functional objects but more about who we imagine ourselves to be.

Experience prototypes are a vital device for addressing these design challenges. They demonstrate what it is like to actually use a product in a given situation and provide findings that can help develop a product's experiential qualities through an iterative prototyping process.

Experience prototypes, often fully working and robust enough for trialling with end-users over periods of time, can play a vital role not just at the concept stage but throughout the design process. They enable the design team, users and clients alike to engage with a concept and prompt vital dialogue between all the stakeholders. They help to ensure a streamlined development process that avoids costly mistakes or delays in bringing a product to market.

Having a working interactive model (experience prototype) enables the design team to learn from a simulation of the proposed product's use in a variety of different contexts and to gain valuable insights into what the experience might be like for users. This can help them to uncover unanticipated issues and needs, and assess the utility and other aspects of the product proposal.

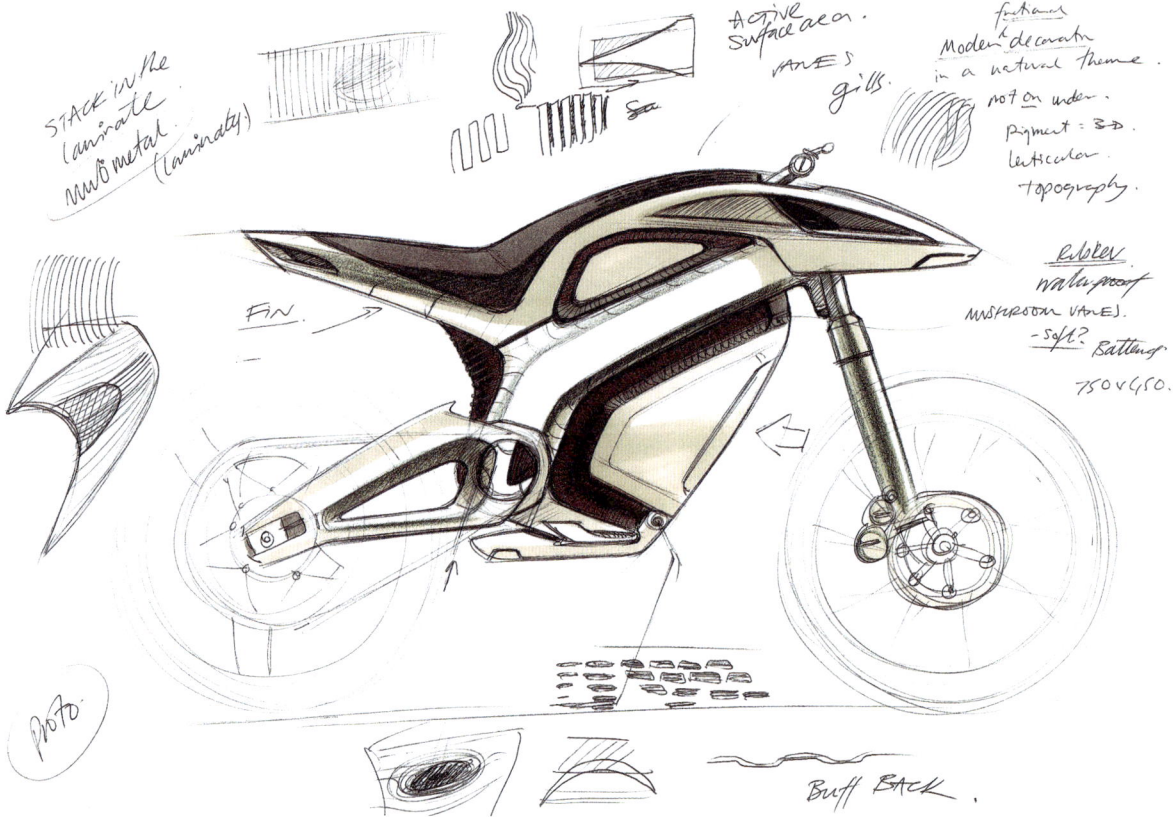

Fig. 8
Development sketch for the ENV motorcycle by Seymourpowell with Intelligent Energy.

APPEARANCE MODELS

An appearance model closely simulates the look of a production product. It is used to communicate a design to clients and users. An appearance model typically does not function in the way a production product would, if it has any functionality at all. Normally internal components do not exist, while all moving parts are fixed in their most preferred or typical position. Sometimes the product is constructed from particular materials so that its weight is represented accurately. This helps assess how a proposed product (for example, a laptop computer) will fit into its environments of use as well as determining whether its physical characteristics are appropriate for the product's purpose.

While large-scale products such as cars, bikes and boats are often presented as a 1:4 or 1:10 scale models, most consumer product appearance models are life-size. The primary purpose of such a model is to evaluate a design's aesthetics and ergonomics, and convey detailed finishes, textures and colours – commonly described as CMF (colour, material and finish). It allows designers to explore the basic size, look and feel of a product.

Appearance models are often hand-carved, sculpted or machined from a solid block of inexpensive material such as foam, plastic, wood or clay, and subsequently finished and painted to look like the desired end product. Due to the materials used, these models are not especially durable and should be handled with care. Appearance models are used for market research, exhibition display, executive review and approval, and product literature photographs. Due to their delicate nature they are not commonly used for interaction and handling by representative users or consumers.

In the past, appearance prototypes were commonly made out of clay; they are subsequently still referred to as clay models in some design agencies and companies. Car companies still do use 'clay' – a form of industrial plasticine – for sketch and appearance models, as it is a malleable material that can be

Chapter 5 Making

Fig. 9
Full-scale appearance models created as physical prototypes, in clay, to give an idea of vehicle mass and rider ergonomics, and as a CAD model, which demonstrates the soft curves of the bodywork.

easily sculpted, when determining a product's form, proportions and surface tension. Its continued use helps to retain the strong practical, aesthetic and conceptual connections between sculpture and design.

The ongoing development of digital visualization software and augmented- and virtual-reality technology has opened up the potential to evaluate product aesthetics without the need for the production of physical appearance models until later on in the design process, reducing costs and speeding up development times.

EMPATHY TOOLS

Over the last decade, the design industry and society as a whole have begun to treat older people and individuals with disabilities differently, moving away from the outdated perspective of viewing them as special cases, and embracing a new social-equality agenda that aims to integrate them into the mainstream of everyday life through a more inclusive approach to the design of products. In addition, there is an increasing recognition of the importance of considering cultural diversity alongside universal design principles. These welcome changes have been reinforced with the passing of equality and discrimination legislation, to which designers are legally obliged to adhere. By addressing the needs of these users, designers can produce better designs that enhance how a broad range of users, regardless of their background or abilities, experience their product designs, increase their potential customer base, and ensure a more equal and cohesive society. Designers need to be aware that inclusive design is an integrated approach that extends to all stages of the design process; it is not simply a stage that can be bolted on.

By empathizing with users, designers are better able to embed inclusive design within the design process and, as a result, produce mainstream products that are pleasurable, desirable and satisfying to use. Many companies and designers, while agreeing with the basic principles of designing inclusively, pay lip service to the practice. They assume that if a product is deemed easy to use, then they are adequately covering their social responsibilities, or naively assume that it is always possible – or indeed appropriate – to design a product to address the needs of all cultural groups.

To avoid such pitfalls and promote an inclusive design agenda, designers should develop an awareness of the needs of users with varying capabilities through empathy and learn how to accommodate them into the design cycle.

Fig. 10
Tom Bieling of Design Research Lab, Berlin, conducting self-experience research into the way blind and partially sighted people interact with an exhibit.

Chapter 5 Making

This design approach, based on the observations of real user actions and behaviour, is exemplified by the design methodology of IDEO and their use of empathy tools. These tools must also consider cultural contexts to ensure relevant and respectful design practices.

Empathy tools, also known as 'capability simulators', are a commonly used inclusive design research method. They are physical or software devices that designers use to reduce their ability to interact with a product, and therefore gain a deeper sense of the experience had by users with disabilities or certain conditions and a better understanding of their needs and desires. Weighted gloves or sports braces, for example, reproduce a loss of dexterity or movement, or spectacles can be smeared with grease to simulate a loss of vision.

These tools can be used throughout the design process to help simulate the physical and cognitive challenges that a design needs to address. However, no capability simulation device can ever truly reproduce what it is like to live with a particular capability reduction on an everyday basis, and they should never be considered a replacement for involving a diverse range of real users, including those from various cultural backgrounds, in developing, designing and evaluating a product.

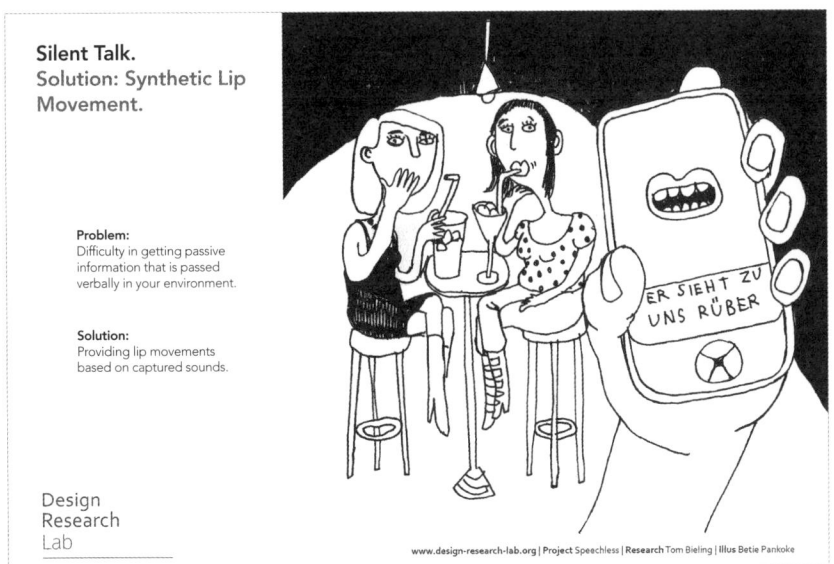

Fig. 11
Product concepts generated by Bieling's research, with a group of hearing-impaired people, into problems surrounding deaf communications. He believes that such products should not only address the needs of a niche group but also widen the field for potential use by a larger group of people in different contexts.

Fig. 12
Exploring the physical constraints and challenges of a problem through bodystorming.

BODYSTORMING

Bodystorming is a useful method that supports empathic working, idea generation and prototyping. It involves physically experiencing a situation in order to derive new ideas, and is especially useful when trying to resolve social and spatial design problems.

In this technique the design team imagines what it would be like if a concept existed and then sets up a scenario and acts out roles, with or without appropriate props, focusing on the intuitive responses prompted by such physical enactment. This technique has been criticized by some design researchers due to the fact that it is often carried out by designers rather than the potential users of the final product, and therefore is not a genuine user-centred approach.

Bodystorming helps design teams to generate and quickly test a range of context- and behaviour-driven concepts, such as eating on an aeroplane, selecting a radio station in a car or ordering a meal in a drive-through restaurant. It has also been successfully used to shed light on design projects such as airline passenger seat design and layout. The goal of the technique is to prompt new ideas – perhaps some that are unexpected – via the physical experience of a situation. A bodystorming session can take anything from ten minutes to one hour, depending on the complexity of the situation and the roles being played out. It can involve one or two individuals, or small or large groups of participants.

Bodystorming can help to generate ideas that might not be realizable by conventional methods such as sketching or model making. It helps create empathy in the context of possible solutions for prototyping. If you get fixated in your idea generation phase, bodystorming can help get you thinking about alternative ideas. It can also be extremely useful in the context of prototyping spatial design concepts, as it focuses on the physical sensations surrounding a design problem.

Bodystorming is a relatively simple method to employ. The aim is to physically 'act out' the design proposition. First, create a list of tasks that you wish to test during the session. Next, go through the list of tasks one at a time. As you are working through each task, verbalize what you are experiencing (e.g. challenges, surprises, other interesting observations). At the same time, ask the other members of the design team to make observations and take notes of what they see and hear.

At the end of the session the design team should answer the following questions:

1. What did you learn from the bodystorming exercise?
2. What surprised you about going through the bodystorming exercise?
3. What did you learn from doing the bodystorming exercise that you couldn't have learned any other way?
4. How can you imagine applying this bodystorming exercise to other design challenges?

RAPID PROTOTYPING

Rapid prototyping is the automatic construction of detailed physical objects from computer data using a range of 3D printing technologies. It first became available during the 1980s. Initially the process could only manufacture prototypes that could serve as a basis for discussion, as they didn't possess the required levels of detail required for serious product evaluation and testing. Today, the range of rapid prototyping technologies has developed enormously and this method can now be used for all stages of the design process, from concepts and appearance models, through to functional prototypes and mock-ups.

Rapid prototyping technologies are also increasingly being used to manufacture production-quality parts in relatively small numbers, and a number of designers, such as Patrick Jouin and Marcel Wanders, have begun to produce one-off and batch-produced products using the technique's unique 'growing' qualities as a viable manufacturing process.

Rapid prototyping is often used to check the design of parts before committing to production tooling. It enables designers to develop a functional prototype model in either short or medium production runs to support a client's development programme. This is necessary when specific functional details and material properties are integral to the finished part design. A range of different materials and 3D printing techniques can be used, and subsequently finished by plating, resin reinforcing and painting. Stereolithography, which creates models in plastic, is arguably the most commonly used process, while other techniques produce models in paper or metal. Designers produce detailed designs on screen, and then output this technical data for production through a form of 3D printing, similar in concept to inkjet printing. Instead of building up text, rapid prototyping actually constructs a 3D object, starting from a computer file by adding one slice on top of another using semi-liquid or powdered material.

Rapid prototyping has become an essential research and development tool in contemporary product design. It enables the quick and accurate creation of tangible physical models in order to verify design details, assembly, aesthetics and ergonomics. The various processes available allow for the layer-by-layer construction of complex models that would be difficult and time-consuming to produce using other fabrication processes, reducing manufacturing time for parts – even the most complex ones – from days, weeks or months to hours.

Fig. 13
Rapid prototyping.

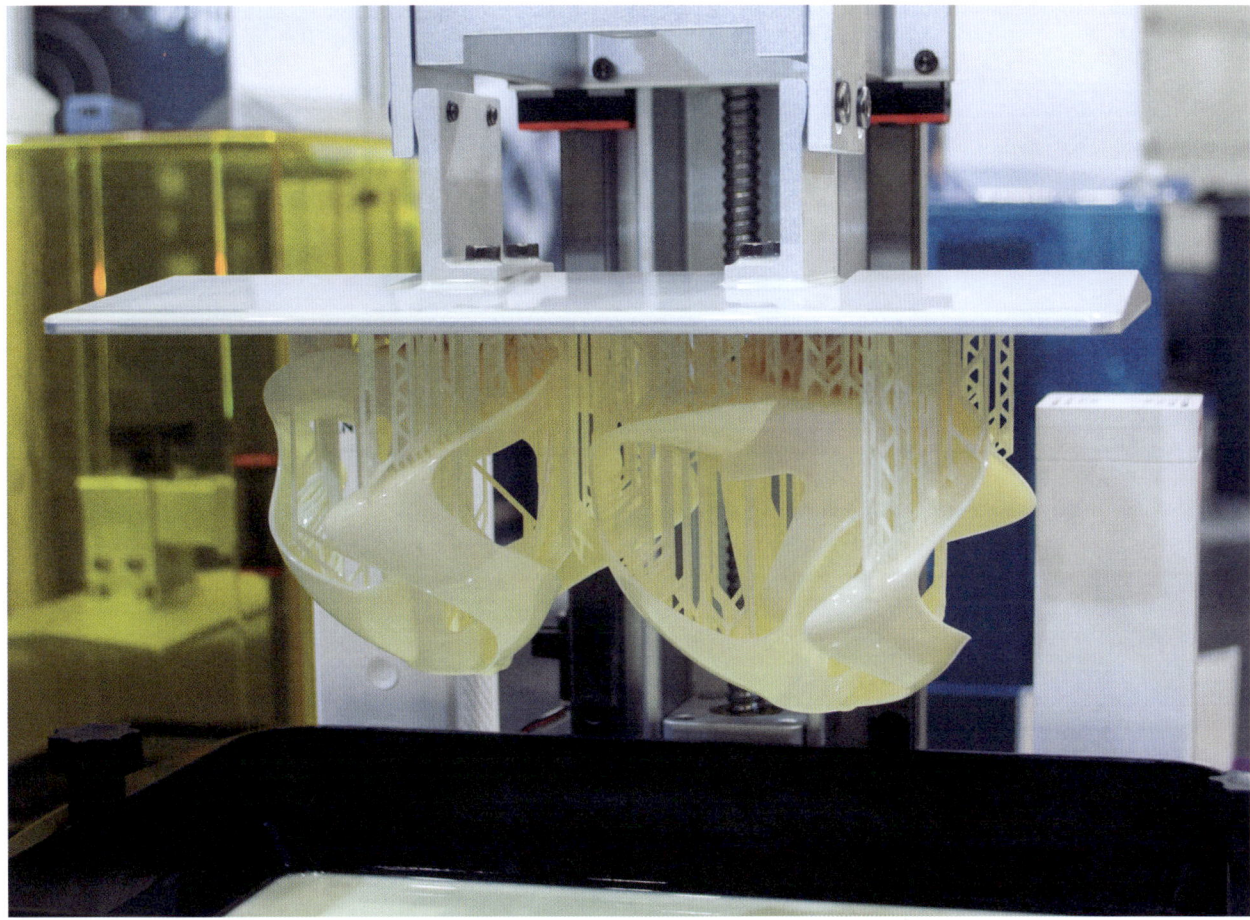

Fig. 14
Design Partners' G-930 Wireless Gaming Headset for Logitech. The CAD models were used to create physical foam models on Design Partners' in-house four- and five-axis CNC (computer numerical control) machines.

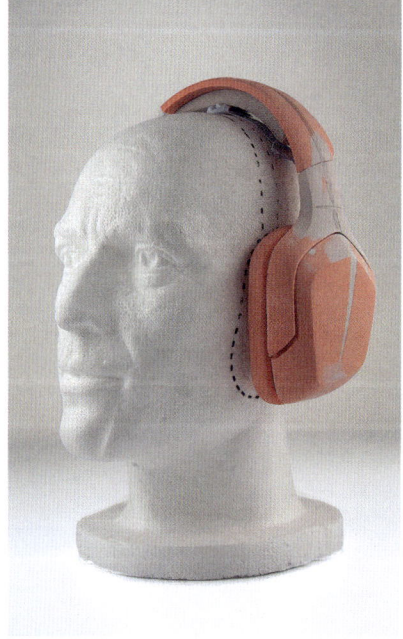

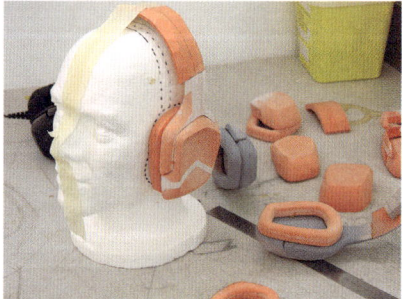

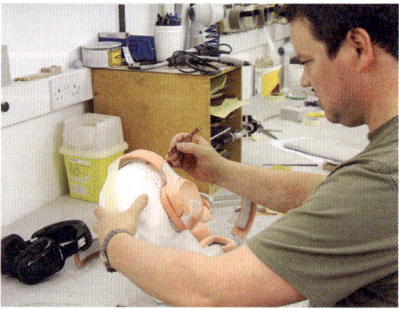

Fig. 15
The Entropia Light by Lionel T. Dean was designed to suit the process of rapid prototyping. Dean set up FutureFactories in 2002 to research the creation of products using 3D printing. The lamp was created in collaboration with Italian lighting design company Kundalini using Selective Laser Sintering, and was 'printed' using the protyping machine as a production tool.

CASE STUDY

4C DESIGN'S NUMNUTS

Introduction
Consumers rightly now demand products that meet strict ethical and humane standards. This project involves Jodie and Andy, who run a sheep farm in eastern Australia, 300 miles inland from Sydney. They have built and run a business producing merino wool and provide it at premium prices by meeting and surpassing stringent animal welfare standards driven by government bodies, buyers and consumers who want ethically sourced products.

Objective
High animal welfare comes at a significant cost in time and money. As a commercial business, farms must weigh up the advantages of welfare, improved health, growth and higher value against increased costs. Several practices common in sheep rearing have been optimized to give lambs a better life. However, humane alternatives to husbandry procedures such as tail docking and castration have not been addressed in almost sixty years. Currently, castration is done with a small rubber ring that severs blood flow and the nerves over the period of one hour, which is a necessary procedure to reduce deadly infections and inbreeding. 4c's objective was to provide a cost-effective and commercially viable solution for offering pain relief.

Methods
The key was to balance both the design and business case by providing a solution that was not only empathetic to the different stakeholders (farmers, vets, pharmaceutical manufacturers, etc.) but also a quality product.

4c worked closely with Moredun Research Institute and CSIRO (the Commonwealth Scientific and Industrial Research Organization) to design various trials in both Scotland and Australia to understand the biological composition of lambs and what practices would work to effectively provide pain relief. They also collaborated with vets, surgeons, needle specialists and anaesthetists in the early days of the project to carry out multiple 'innovation sprints' to quickly conceptualize potential solutions that were trialled on test rigs to understand usability and ergonomics.

This was followed by six rounds of 'wild prototyping' to improve every aspect of the process in order to understand and add business value. The team found ways to reduce repetitive strain injury (RSI) and to develop the product form to suit different-sized hands, which current products ignore. They also looked at the treatment of the lambs during these procedures, which could result in unnecessary bruising and tissue damage.

As the process used a drug and needle system, the design team explored the user experience and built key requirements for the product. To reduce the chance of needle-stick injury, a novel injection mechanism de-skilled the process and forced hands away from the injection zone to add security to the drug delivery operation. At this point, it became clear that there were three products to develop: a castration tool, an injection mechanism and a drug-delivery consumable. This resulted in months of designing, prototyping and testing with farms in Australia. As a small start-up, 4c had to tailor its work to suit existing systems, designing to suit set bottles and filling stations. To understand the final product requirements, 4c carried out systematic design exercises with the end-users and potential suppliers of the product.

Chapter 5 Making

4c also developed an online veterinary portal for vets to sign up to distribute the local anaesthetic, and an online shop for farmers to purchase the hardware. To compare the product with competitor products on the market, 4c built test rigs and designed numerous tests to quantify the improvements in usability, repeatability and durability. 4c launched Numnuts in Australia after finalizing supply chains for the 2019 lambing season. This required a 'feet-on-the-ground' approach: going to farms, speaking to vets, attending tradeshows and taking novel approaches to fundraising for future batches.

Results
To date over one million lambs have received the Numnuts treatment, and the farmers' reviews are wholly positive. Drug companies now sell considerably more local anaesthetic, and vets have a new business model. For farmers, new suppliers are interested in their produce and are willing to pay more. For consumers, people are paying more for sustainable, better-sourced produce. And, most importantly, lambs receive improved welfare.
4cdesign.co.uk/work/numnuts/

Top: Development sketches.
Bottom: Final product.

CASE STUDY

PILI WU'S PLASTIC CERAMICS

Introduction
Developed during the Chinese Song Dynasty, eggshell porcelain or bodiless chinaware is a unique technique which produces remarkably thin and light porcelain wares, which feel almost as thin as eggshell. These porcelain pieces can easily be seen through when held up to the light.

Objective
The fine eggshell porcelain wares are often exclusive products made in limited quantities, since the production processes are very complicated and time-consuming. The HAN Gallery set Pili Wu a brief that asked him to research these complicated processes in order to make a series of fine eggshell porcelain wares that would be much more affordable for mass markets.

Methods
Pili Wu initially conducted a series of web searches and literature reviews into the history of the manufacture of eggshell porcelain. He discovered that, traditionally, eggshell porcelains are first fired at around 800 degrees Celsius, and then treated by very experienced craftsmen to carve out excess surface material to achieve the desired thickness. In the end, the eggshell porcelain pieces are fired again at 1280 degrees to bring about the final products. However, the porcelain wares easily deform and break during the firing processes, due to the insufficient structural force rendered by the extra thin walls. Therefore, decisions had to be made to increase the strength of the porcelain ware walls, while maintaining the eggshell-like thinness. To continue the project, Pili Wu further researched using a combination of methods including literature reviews, web searches and product analysis of existing disposable plastic wares that are commonly used in Taiwanese roadside restaurants. He found that the thin and fragile surfaces could be strengthened by a rib-like structure. The creative link was made by adapting this structure into the designs of the new set of porcelain wares.

Results
After a series of trial and error methods that included porcelain mock-ups, material testing coupled with the designer's intuition, the project has been deemed a huge success. The final mass-produced tableware family includes a fruit bowl, a candleholder, a Chinese bowl, a teapot and a set of three cups. Based on the ancient techniques, the design of Plastic Ceramics brings not only the traditional craft of eggshell porcelain wares but also the elegant lifestyle of the Chinese literati into the contemporary world. The industrial motif stamped on the base of each piece reinterprets the intricate patterns of rare antiques, giving a fresh look to the refined, lightweight porcelain tableware which is now suitable for both Chinese and Western tea-serving rituals and global table landscapes.
piliwu-design.com

Chapter 5 Making

The Pili Wu mass-produced plastic ceramics tableware family includes a fruit bowl, a candleholder, a Chinese bowl, a teapot and a set of three cups, based on ancient techniques developed during the Chinese Song Dynasty. The design of Plastic Ceramics brings not only the traditional craft of eggshell porcelain wares but also the elegant lifestyle of the Chinese literati into the contemporary world.

TUTORIAL

How to conduct experience prototyping

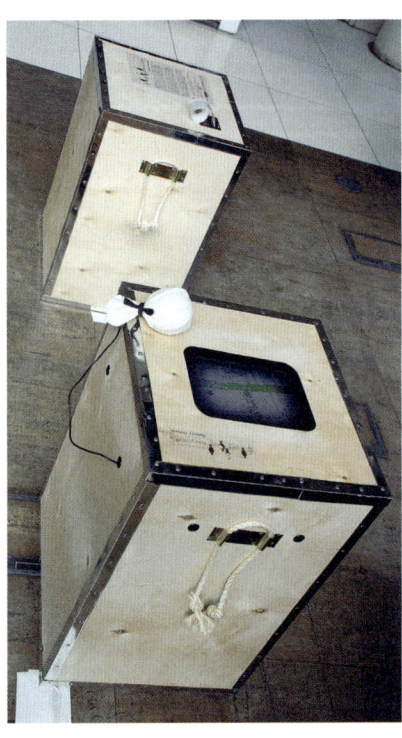

'Sensory Threads', a multi-partner collaboration led by Proboscis, is a project where groups of people can create a collective soundscape of their interactions by carrying wearable sensors. The data is fed to the 'Rumbler', where it can be experienced remotely as vibration, sound and image. The Rumbler allows people to play back the sonic and sensory explorations.

Experience prototyping allows designers, clients or users to experience a proposed product, service or system for themselves rather than witnessing a demonstration of someone else's experience. Experience is, by its very nature, highly subjective; the best way to understand the experiential qualities of using a product, service or system is to actually experience it first-hand. So, just like the rapid prototyping of physical objects or software interfaces, designers can create experience prototypes that enable users to experience a new product, service or system before it actually exists. Experience prototyping is an extremely powerful tool in gaining user insights early in the design and development process.

When creating an experience prototype you should utilize a combination of prototyping methods and materials (e.g. print and paper prototypes, bodystorming, SMS, phone calls and so on). The emphasis should be on speed – prototyping quickly, testing with users, learning, improvising and iterating. It is also vital that all user 'touchpoints' (i.e. the designed elements associated with the proposed product, including interactive internet features, printed documents, physical devices, retail outlets and call centres that the user might come into contact with) are represented as well as possible so that participants can feel immersed in the experience of the proposed product, service or system. Therefore, a rigorous 'end-to-end' experience prototype for a new bike-hire service, for example, may well include paper prototypes, screen-based interactive simulations (perhaps with mock-up hardware), SMS service, voicemail service and in-store mock-up environments that enable users to experience the complete service.

User insights gained from experience prototyping sessions can be extremely revealing and inform product, service or system propositions and other design decisions. The results of a well-executed end-to-end session might include:

— Qualitative feedback about the users' experiences.
— Reliable quantitative user interaction timings.
— Insights as to the best sequences of information presentation.
— Assessment of user comprehension.
— Identification of user errors and any underlying reasons for them.
— Identification of opportunities to improve the design.

Chapter 5 Making

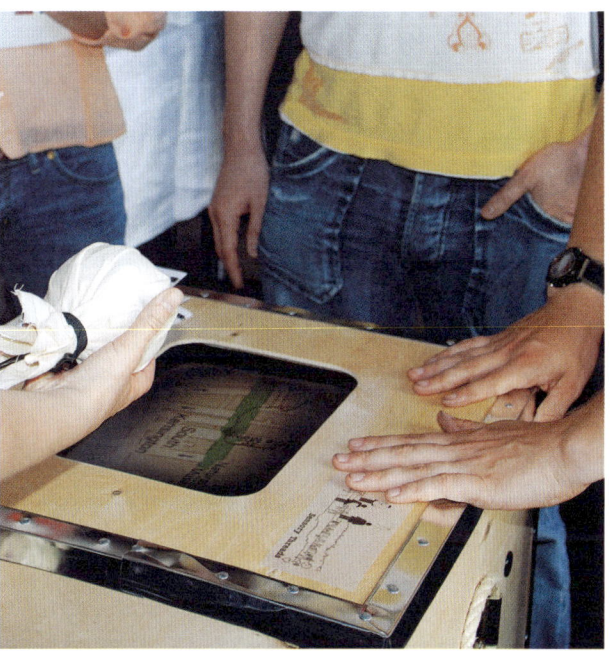

'Sensory Threads' uses music and vibration to alert our consciousness to barely perceptible changes in the environment. Variations in the soundscape reflect changes in the wearers' interactions with each other and with the environment around them.

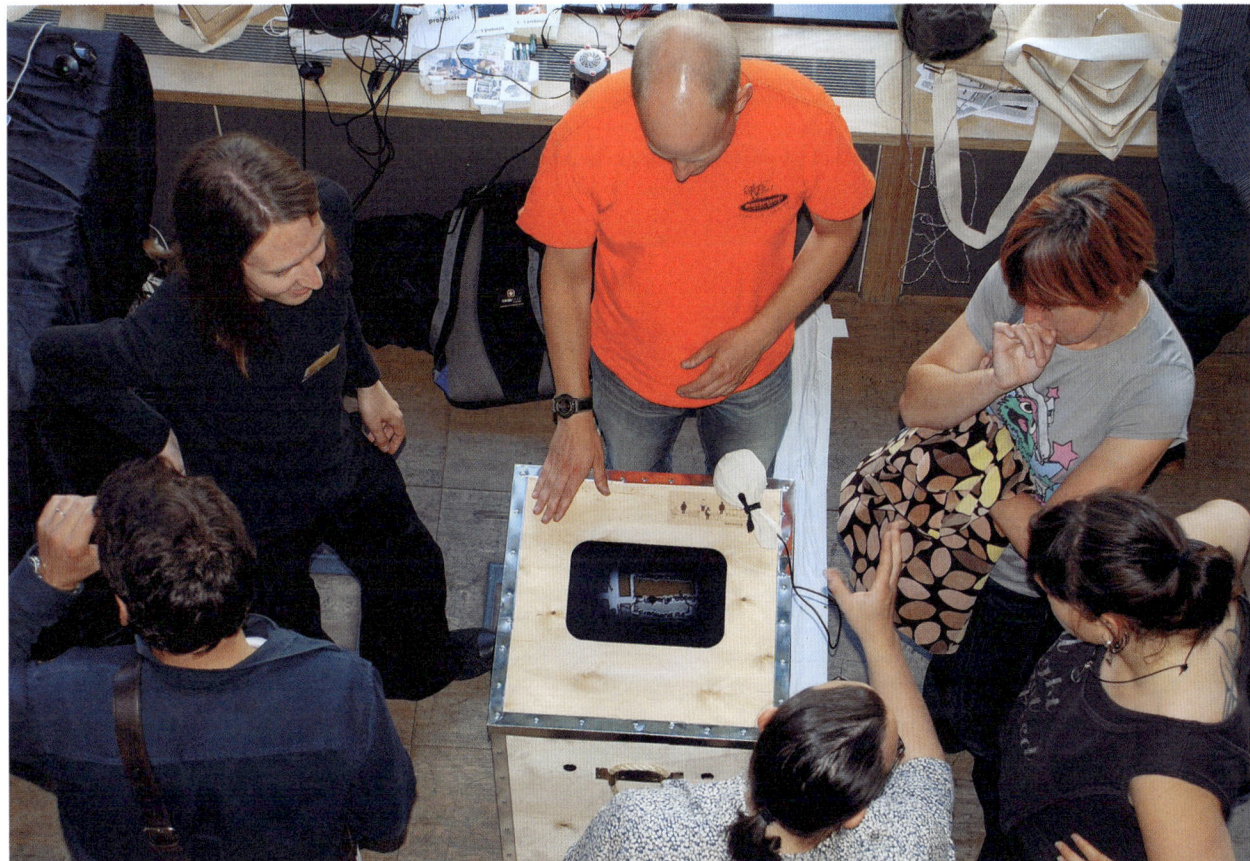

TUTORIAL

How to do quick-and-dirty prototyping

Quick-and-dirty prototypes are usually constructed using inexpensive materials such as paper, card, glue, sticky tape, and polystyrene foam.

Designers need results rapidly, and as a result quick-and-dirty prototyping has become an increasingly popular approach to design development. Quick-and-dirty prototypes should make your thinking tangible, giving shape to your ideas. They provide invaluable insights into how a product concept will be used, and because of their 'rough and ready' quality, people are not afraid to criticize these test models.

When considering making a quick-and-dirty prototype, you should start by determining which aspect of the user experience you want to test, and choose an appropriate 'good enough' representation to test it. This will vary according to the stage of development your project is at. For example, you might test the initial concepts behind a new chair design by building a series of skeletal structures that help define the ergonomic layout and structural requirements that need to be met. You might then build 'work-like' models that test mechanisms, and separate 'look-like' models that focus on aesthetics. By doing this in teams, you can foster a collective understanding of what you are trying to achieve, and produce a far larger number of models than would be possible working independently.

You should use whatever materials are readily available – remember that this form of prototyping only needs to be good enough to test the issue at hand. Any additional refinement can actually be detrimental to the rapid iteration of design concepts. Typically you should have paper, card, foam board and hot glue at hand, but you might need to access a DIY store, depending on the nature of your designs.

As mentioned earlier, quick-and-dirty models are produced on an appropriate scale – smaller than actual size for large products, actual size for products such as mobile phones or hand-held devices, or larger than actual size for very small products. The scale you select will also depend on the stage of the design development, the time you have and the needs of the client and other stakeholders. In the early stages of a design project, for example, many smaller-scale models might be more appropriate.

A quick-and-dirty prototype will necessarily cut corners. The fine details of the design proposition are not needed – the emphasis is on getting approximate and quick results and/or feedback. The quick-and-dirty method of model making is a particularly effective tool in rapidly communicating design ideas to diverse groups of stakeholders. You should also involve potential users in the design, build and evaluation of your prototypes.

Chapter 5 Making

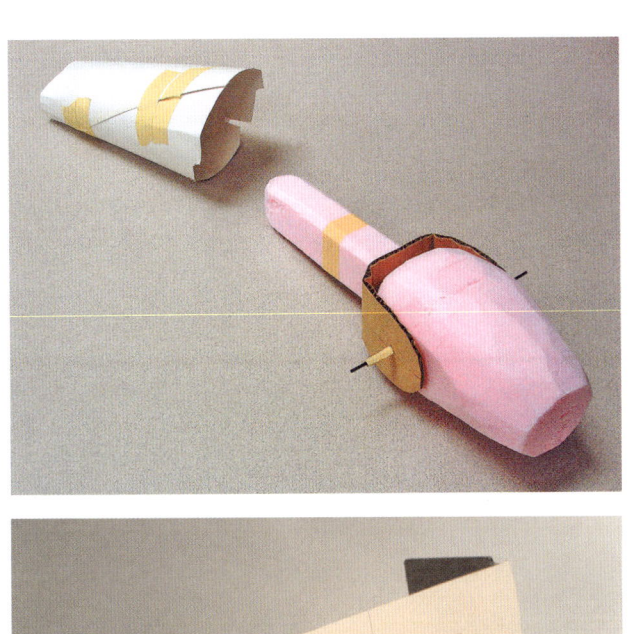
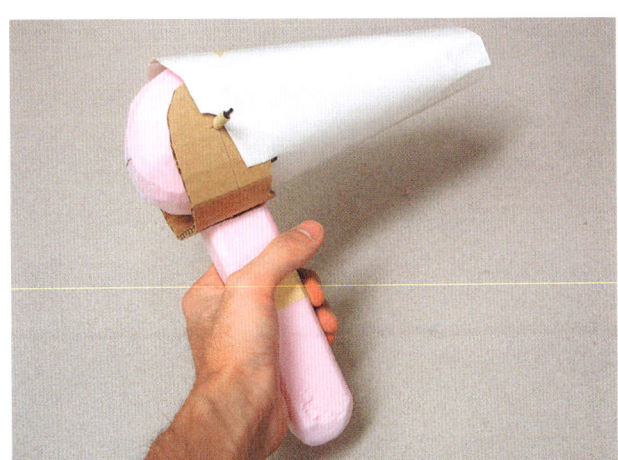
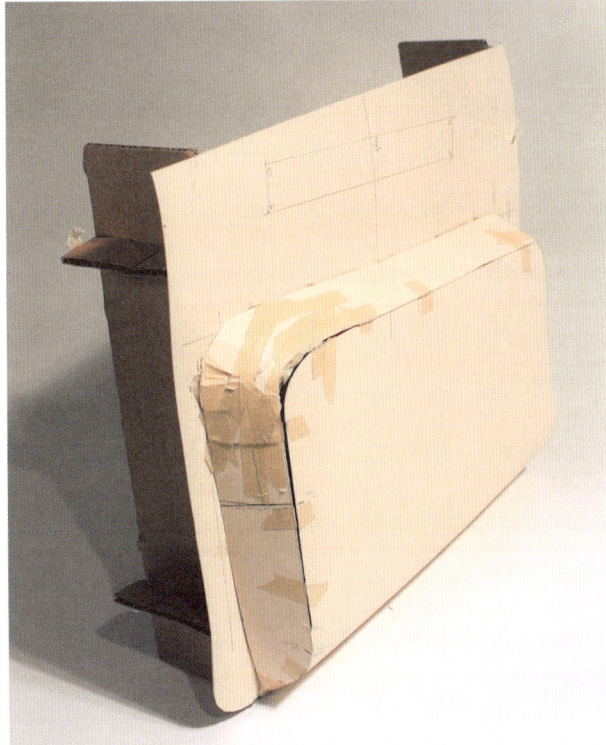

Shown here are prototypes for a folding hair dryer, using foam and paper, and a laptop bag, using cardboard with pencilled details.

6 TESTING

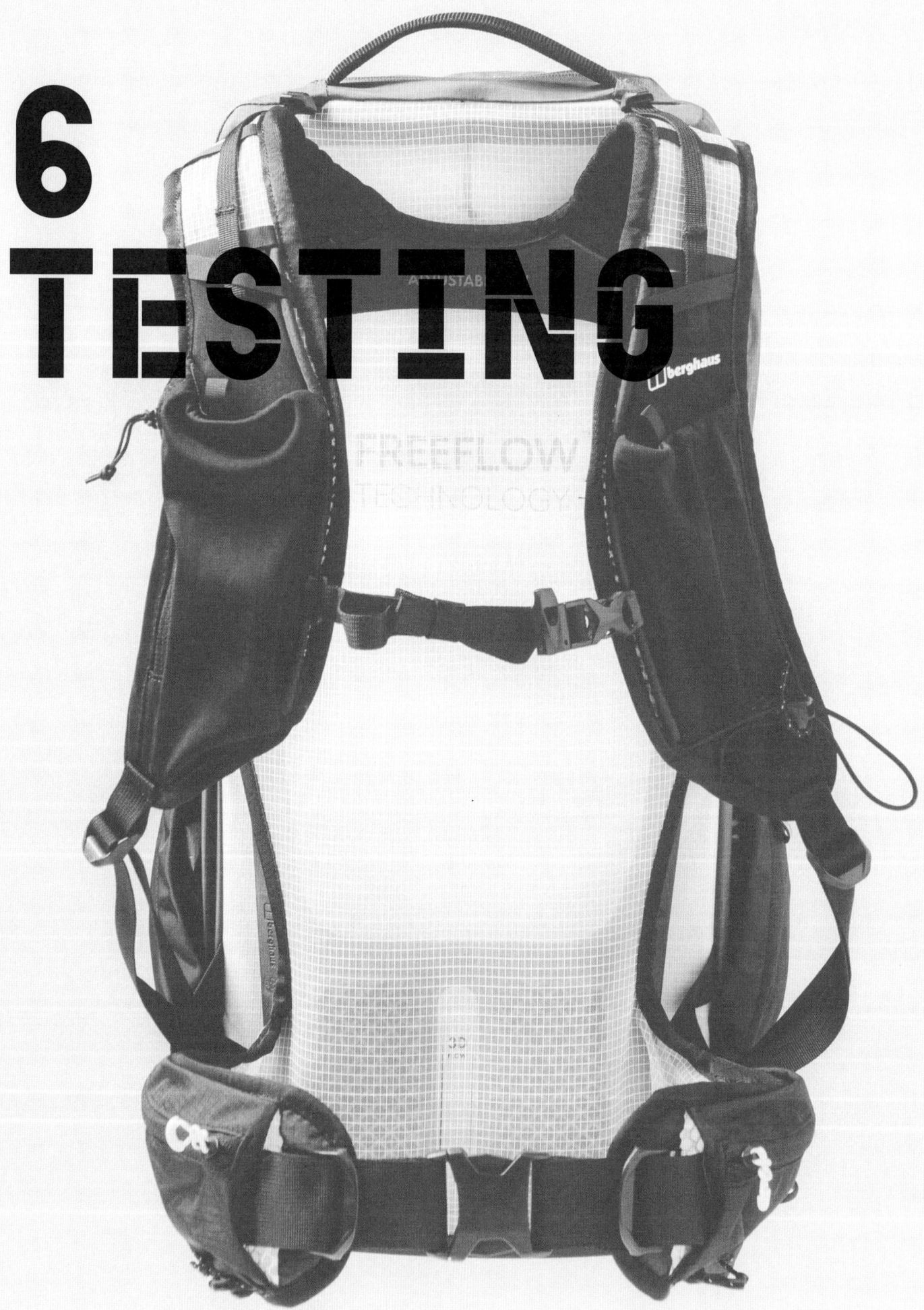

Chapter 6 Testing

Extensive testing is a vital stage in developing and resolving a product. Designers need to test their research concepts and prototypes with end-users over periods of time to enable the design team, users and clients alike to engage with a concept and prompt vital dialogue between all the stakeholders. Testing facilitates informed decision-making, and helps ensure a streamlined development process that avoids costly mistakes or delays in bringing a product to market. This chapter introduces a variety of techniques for testing your designs, including user trials, building test rigs and safety testing.

SCENARIO TESTING

Scenarios can help product designers communicate and evaluate design proposals within their intended context. Design is a form of authoring practice, where designers can craft 'design fictions' that propose possible futures. By devising a scenario carefully with characters, narrative and context, designers can evaluate whether their design ideas will work for their intended users.

Scenario testing involves the creation of future scenarios using media such as storyboards, texts, photography, film and plays to present a product, service or concept, and then asking users to provide evaluative feedback on them. The more appropriate and convincing the presentation format, narrative and scenario, the more likely it is that the feedback will provide useful insights into a product's functional, aesthetic and behavioural qualities.

Designers require a test arena for developing and evaluating sociocultural trends and narratives to help identify emerging values and needs. Scenarios are sketch stories that provide a clear context and set of criteria that designers can use to find solutions for future problems. They connect research to appropriate design solutions. Scenarios are based on in-depth research findings, from ethnographic studies and interviews with stakeholders to analysis and forecasts from subject experts. These research findings then provide the basis for imaginary scenario characters. Because most products have a variety of different users, each with their own personal, professional and social concerns, you should typically create at least three different scenarios constructed around different characters to cover the scope of product interactions.

A set of scenarios tells you why your users need your design, what the users need the design to do, and how they need your design to do it. Scenarios aim to tell designers how users will behave in the future, and use as yet uncreated designs and services. They promote a holistic approach to design, and help prevent designers from making design decisions based on their personal experience or preconceptions.

New products create new behaviours, and the use of scenarios can enable a design team to test new product concepts within a carefully constructed 'future context', exploring and evaluating the experiential opportunities and shifting behaviours. Scenarios are especially useful during the early stages of a product's development, when the design team can communicate, test and evaluate the validity of an early concept design to a client and the appropriate stakeholders.

Scenarios are also commonly used to provoke discussion, creating propositions that can be described as 'discursive'. While traditional 'good design' often professed to be unobtrusive, intuitive, invisible and something that does not make the user think too much, discursive design instead seeks to encourage self-reflection, stimulate the audience's imagination and provoke considered debate. A discursive product scenario can be thought of as a tool for thinking, with a conceptual design given form and function by the designer so that it can communicate ideas, which can be tested out through discussion with users and communities.

Fig. 1
Manufacturers extensively test their products through user trials and expert test drivers to ensure the required physical, tactile and dynamic qualities are met.

Fig. 2
(Opposite) Testing 'Intimate Mobiles' by Fabien Hemmert at Design Research Lab in association with Deutsche Telekom Laboratories. These experience prototypes allow for near-body telepresence in mobile telecommunication, expressing airflow, moisture and tightness. All means are explored through mobile phone-shaped and -sized boxes, which are equipped with the necessary actuators.

USER TRIALS

User trials are an effective experimental method, where a group of users test versions of products under controlled conditions. They are often carried out as part of initial research to evaluate existing products, or when a complete product is to be evaluated. Rough-and-ready prototypes may also be used, but they must be robust enough to ensure they don't degrade over the testing process and provide flawed data. User trials are often used before a design has been signed off for manufacture, and are commonly used on pre-production prototypes. They are often used as a cost-effective way of evaluating products compared to more extensive field trials, which commonly take place when a more complete product is to be tested and evaluated prior to launching into the marketplace. These studies help to inform design researchers' understanding of the specific components and actions of the user during a set task.

Setting up an effective user trial is primarily about creating an environment that enables the interaction between a product and a user to be systematically examined and measured. A representative sample of 'real' users is recruited and they are provided with a series of tasks to undertake within a set timetable. Information relating to the time spent completing a task, or the number and types of usage errors made, is employed to compare different versions of the same product and/or user interface.

User trials commonly use video recording to capture observations, and may also rely on a team of trained observers in a controlled 'laboratory' setting to identify and record specific issues that users encounter when undertaking the tasks. Some researchers feel, though, that this level of laboratory-type observation can colour the findings – that the presence of an observer can affect the behaviour of a user – and they may choose to avoid this in order to encourage a more 'natural' interaction with the product under test by removing the possibility of users' actions and behaviour being affected by observers being present. In addition, when selecting participants the design researcher should ascertain if they have prior knowledge or experience of using a similar product.

Once the users have completed the trial they are interviewed about any difficulties they encountered or observed during the tasks. These interviews, together with a detailed analysis of the tasks captured on film, provide invaluable objective and subjective insights that can enable designers to improve the ease of use, functionality and intuitiveness of the product being developed. Videos also provide an ideal vehicle for communicating issues to the wider product development team, client and stakeholders.

While user trials are a commonly used research technique they are not without their faults. It should be noted that the research findings provided by user trials are heavily reliant on how well the users have understood the purpose and procedure of the user trial and their ability to communicate their feedback to the evaluator(s). It can also be difficult to source the required types of representative participants for user trial, and once a design researcher has assembled a user trial group, they must ensure that they don't become overly reliant on this group of individuals. Unless observation of the effects of cumulative learning is a specific aim of the research, the user groups should be changed periodically to avoid over-familiarity with the product and the testing process.

Chapter 6 Testing

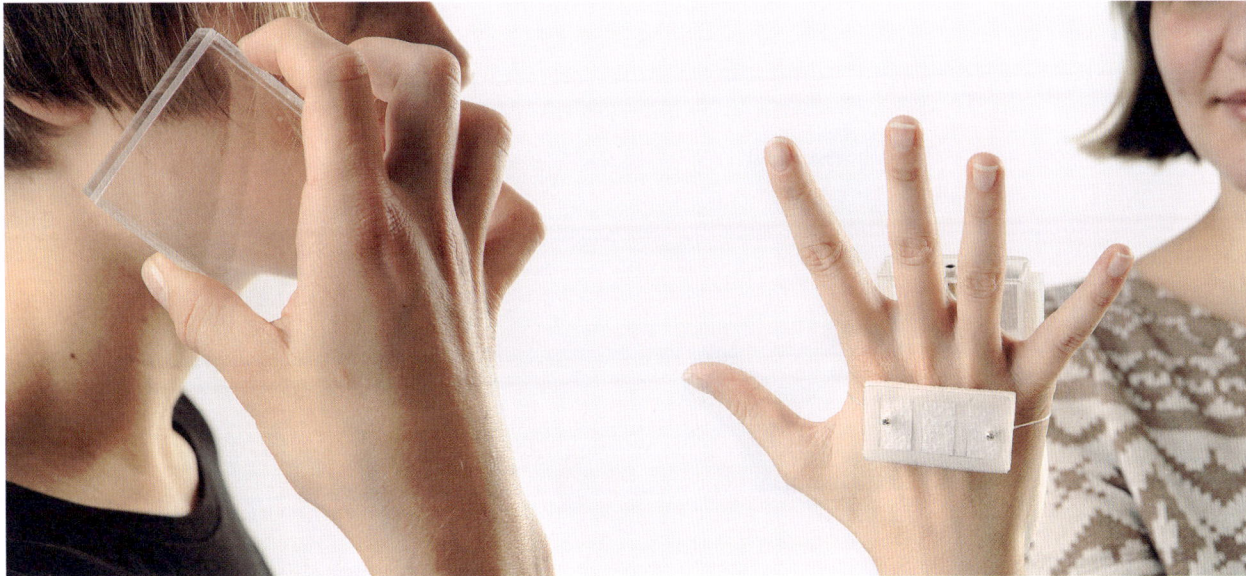

PRODUCT USABILITY TESTING

Product usability testing is a technique used to evaluate a product by testing it on its intended user group, which usually focuses on measuring the product's ease of use and its capacity to meet its designed purpose. Usability testing typically measures the ease of use of specific products such as consumer goods, websites, and digital applications, interfaces and devices. Usability testing is a vital part of the design development process, enabling designers to verify whether a product is achieving its human interaction objectives and goals.

Product usability testing involves representative users being asked to test the product in a realistic environment. By observing the user's behaviour, emotions and difficulties, designers identify attributes and qualities that require improvement. This type of testing also provides designers with the opportunity to see if users interact with a product in unanticipated ways, which could, if harnessed, contribute to the improvement of the design. In the context of digital products, usability testing might include techniques such as remote testing, A/B testing and clickstream analysis to gather insights that are particularly relevant for online interfaces and applications.

Expert reviewers are often asked to undertake initial usability testing and the benchmarking of design concepts against previous products or competitor offerings, as their in-depth understanding of a product type or activity can enable them to identify major usability issues early in the design process. Designers can then avoid pursuing any costly dead ends. Usability testing is especially helpful for evaluating initial design concepts, as designers may test different elements and/or product interactions of a concept, making several small models of each component of the product/system.

Usability prototypes used in the industry range from rough-and-ready prototypes and mock-ups to paper models, visualizations and storyboards. In the realm of digital design, prototypes might also include interactive wireframes and high-fidelity simulations that allow users to engage with the product dynamically.

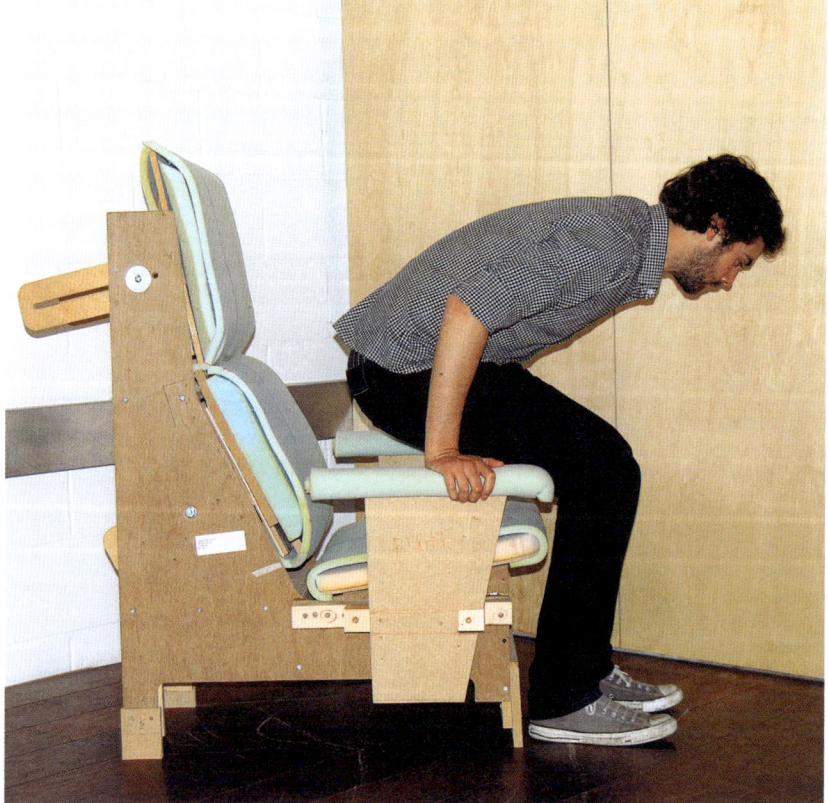

Fig. 3
Testing the ergonomics of an early MDF and cardboard rig of PearsonLloyd's hospital chair.

Chapter 6 Testing

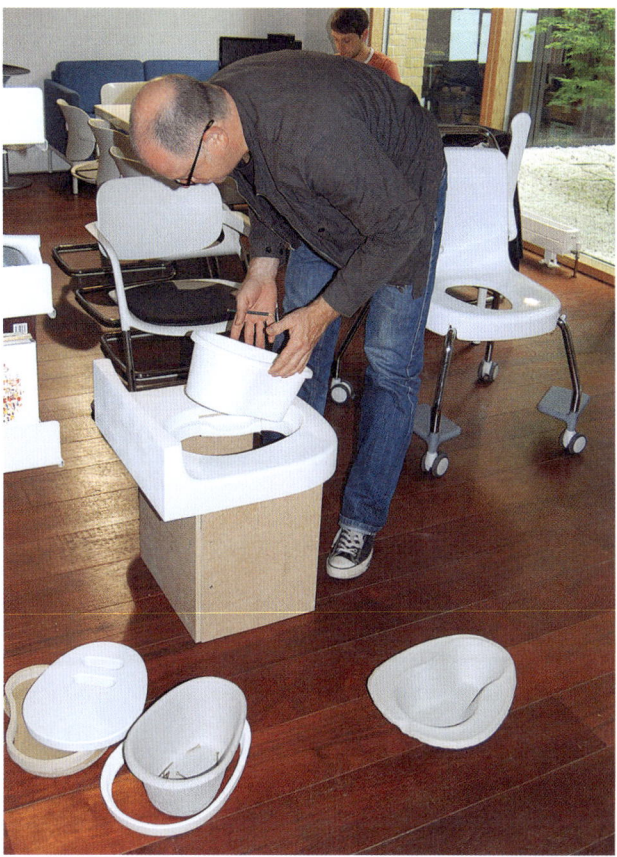
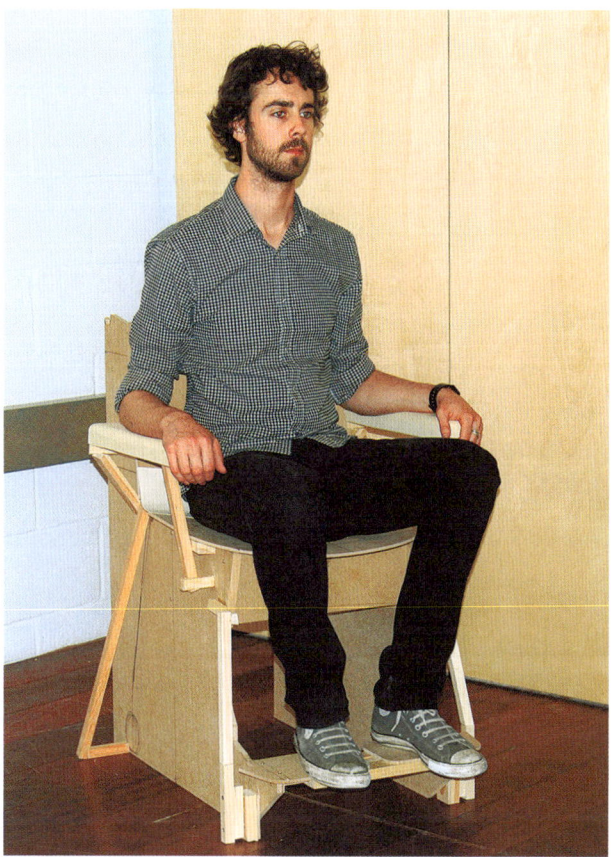

Metric analysis

While conducting product usability tests, designers must determine and then use a set of metrics to identify what it is they are going to measure. These metrics are often variable, and change in conjunction with the scope and goals of the design project. Qualitative design issues such as user satisfaction are often tested alongside more functional usability tasks. These metrics are then measured, producing data such as the percentage of users that completed a task, how long it took the sample group to complete the tasks, the ratio of success to failure to complete them, and the number of times users appeared frustrated. The ultimate goal of analysing these metrics is to find/create a prototype design that users employ – and like – to successfully perform given tasks and interactions.

Rapid iterative testing

Rapid iterative testing and evaluation (RITE) is a form of product usability testing that encourages testers to 'think aloud', enabling the observing design team to step in and change the user interface of a product, interface or service once a problem has been identified and a rapid solution has been devised. The RITE technique is arguably less methodologically robust than traditional testing, but it does dramatically reduce development time, and is commonly used in the software development field.

Remote testing

Remote testing is a usability method that allows users to interact with a digital product from their own location, employing their devices in their natural environments. This approach involves screen-sharing tools, video conferencing or specialized software to observe users as they engage with the product. Remote testing provides several advantages, including access to a geographically diverse user base and increased flexibility in scheduling

Fig. 4
PearsonLloyd explored a range of materials and forms when developing their user-centred commode.

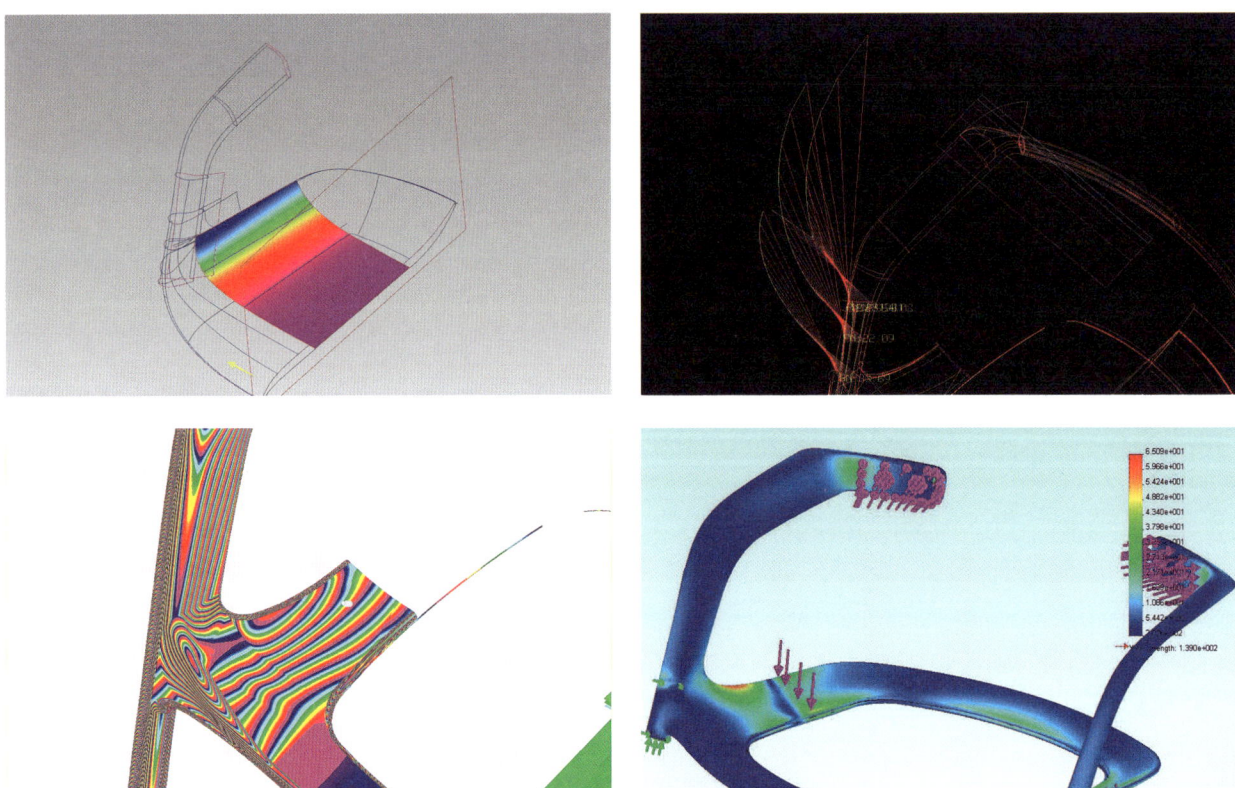

Fig. 5
Graphic CAD images demonstrating how a chair will perform under loading; from top left in a clockwise direction, they show the surface analysis of various components; the pressure each curve exerts on the model; an FEA (finite element analysis) stress analysis to understand impact on sections of the chair; an analysis of the surface using zebra curves.

sessions. Additionally, it minimizes the influence of a researcher by allowing users to perform tasks in a more authentic context, thus yielding valuable insights into real-world user behaviours and preferences. This method is particularly useful for gathering qualitative feedback and understanding user experiences without geographical constraints.

A/B testing

A/B testing, also known as split testing, is a quantitative usability method that compares two or more variations of a digital product to determine which one performs better in achieving specific goals, such as increasing click-through rates or enhancing user engagement. In an A/B test, users are randomly assigned to different versions of the product and their interactions and outcomes are measured. By analysing the performance data, designers can make informed decisions about which design elements resonate more with users. This method is particularly effective for optimizing user interfaces, determining effective content placement and enhancing overall user experience based on empirical evidence rather than assumptions.

Clickstream analysis

Clickstream analysis is a method that tracks and analyses the path users take while interacting with a digital product, capturing every click, scroll and interaction as they navigate through web pages or applications. By collecting this data, designers can gain insights into behaviour patterns, identifying common pathways, bottlenecks and points of friction within the user experience. This quantitative approach helps to highlight where users drop off the journey, which features are most frequently accessed and how effectively information is conveyed. Clickstream analysis is instrumental in uncovering unintended navigation issues and assists designers in refining user flows, ultimately leading to a more intuitive and satisfying user experience.

Chapter 6 Testing

MATERIAL TESTING

When designing a product you need to know if it will cope with every anticipated use in every expected environment. You need to know what materials it should be constructed from, whether it can be manufactured efficiently, and whether it will function at all extremes of tolerances. Selecting the right materials will be one of the most important product development decisions you will make. Failures due to poor material selection are all too common in some areas of product design, and when the wrong choice of material is made, the unanticipated costs in terms of project delays, warranty claims or product recalls can be significant.

Failures due to poor material selection are all too common in some areas of product design. For example, in the plastics industry, it is vital that you understand the properties of plastic, research the material you are working with, and undertake some application-relevant testing. The sort of questions you will be seeking to answer when you are considering which material to use are: does this material have temperature- and/or time-dependent properties? Will it age physically? Will it be susceptible to chemical or environmental attack? Will it be susceptible to weathering? Will the construction result in stress problems from moulding, welding or filling? Don't rely on your experience alone or on the word of a supplier – ensure you test your designs fully.

These technical evaluations are undertaken through the use of test rigs, usually constructed in performance-testing laboratories. A test rig is the term commonly used to describe a full-size or scale model that replicates a mechanical action or enables a design's physical properties (such as its strength, stiffness, comfort or durability) to be tested. While computer modelling and analysis techniques provide very important insights into the

Fig. 6
Testing a chair using heavy weights to pound, bend and flex the chair through thousands of cycles to see how long it will take to fail.

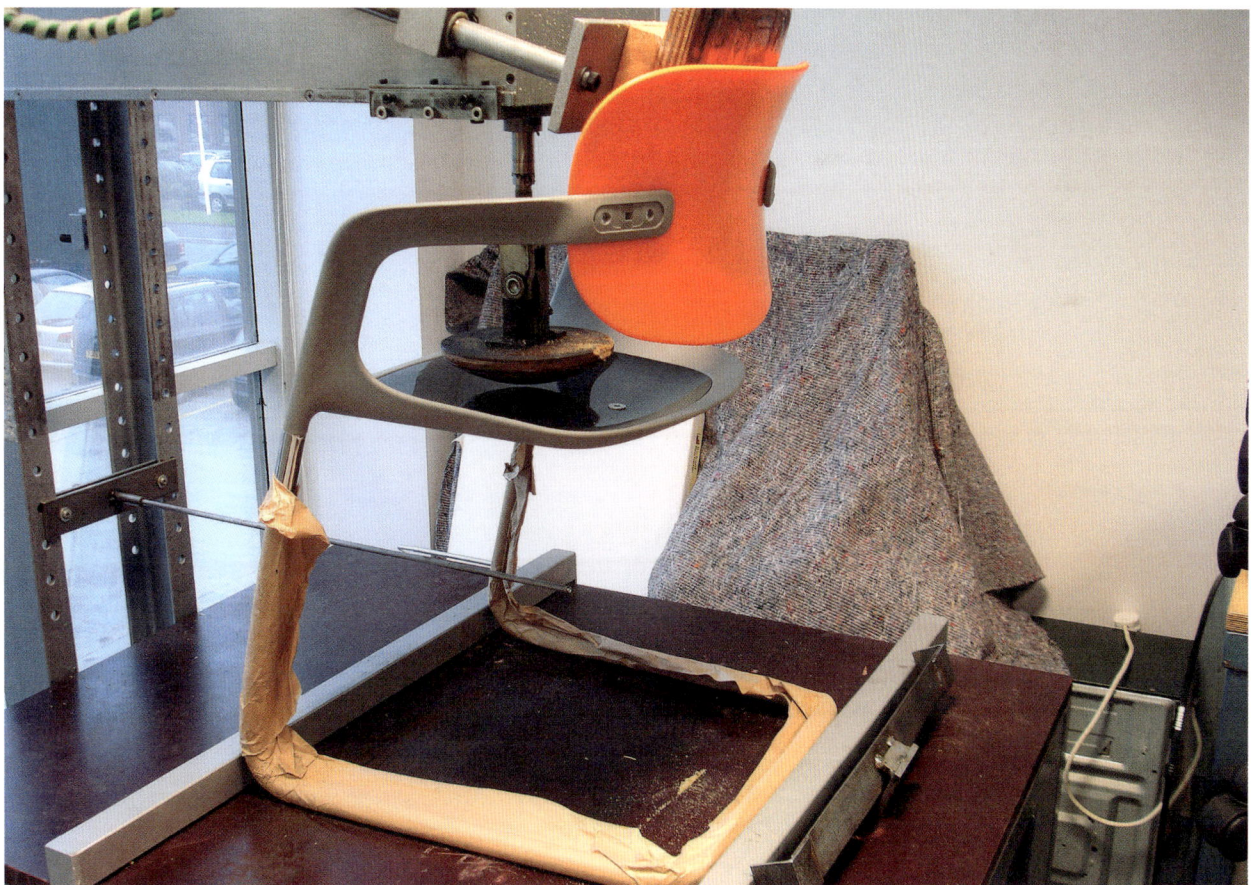

Fig. 7
Selection of international safety standard marks. (Top to bottom) Conformité Européenne, Canadian Standards Association, British Standards Institution, BEAB (British Electrotechnical Approvals Board), Germany's TÜV mark and America's UL (Underwriters Laboratories Ltd).

Figs 8 & 9
(Opposite) Cars are subjected to the most rigorous tests. The images opposite show seatbelt testing using crash test dummies, and testing tyre skids in a controlled environment.

likely performance of a product and its constituent components, even the best of them are based on assumptions and approximations of actual product behaviour. The extensive practical testing and evaluation of prototypes through the use of test rigs is the only real way to validate a physical design.

An example of this is drop testing. This is the term used to describe the technique for measuring the durability of a part or material by subjecting it to a free fall, from a predetermined height onto a surface, under prescribed conditions. This is a compulsory test for electronic goods, and helps ensure that components, fastenings and tolerances are appropriately robust.

The practical physical testing of products is a crucial aspect of a company's design, development and risk management strategy. Designers work closely with engineers during product testing to ensure that lessons learnt are fed back directly to the development team, designing in reliability and robustness. This is done through the use of structured risk analysis tools, such as Failure Modes and Effects Analysis (FMEA), which enable the testing team to identify, quantify and mitigate the specific risks associated with product assemblies or individual components, and their proposed methods of manufacture. FMEA determines the location and nature of a failure, and whether a material defect contributed to the failure, helping the design team to understand and prevent future failures.

Such analysis helps inform the development of a meticulously planned, conducted and documented testing regime, which can be used to investigate alternative design solutions, prove a particular design solution principle, investigate alternative design solutions or carry out specific robustness tests. During the design of most products and their constituent components, the expected physical stresses and loads the product will need to endure are determined, and these are used to design the required components and inform material selection.

Devising and running a well-controlled laboratory test rig evaluation, rather than relying on field evaluation, can lead to a much better understanding of the strength of a set of components, and can enable the iterative development of improved designs more rapidly than relying on field evaluation.

SAFETY TESTING

Product safety is a primary concern when designing, testing or, indeed, purchasing a product. A product may be unsafe due to a fault in either the manufacturing or the design process. A fault in the manufacturing process will lead to a product not functioning as intended. For example, the locks on a folding baby buggy may fail to engage properly when the buggy is unfolded, causing it to collapse. Such safety faults can be identified through extensive testing, and problems with the production quality, strength or tolerance of components can then be resolved through production design changes.

A product may, however, have been manufactured as intended and function properly but be unsafe due to a design defect. For example, a folding chair may unlock unexpectedly when the user tries to move it, trapping their fingers. This type of problem is usually due to a product being designed to meet a set of criteria that fail to accurately reflect real-world conditions. This will then have been compounded by the failure to pick up the problem in testing due to a lack of insufficient material and/or user testing.

Safety testing is increasingly a key issue in determining consumer purchasing decisions. A prime example of high-profile safety testing undertaken today is the European New Car Assessment Programme (Euro NCAP). By law, all new car models must pass certain safety tests before they are sold. However, while this legislation provides a minimum statutory standard of safety for new

Chapter 6 Testing

cars, it is the aim of the Euro NCAP tests to encourage manufacturers to exceed these minimum requirements.

Current testing evaluates adult and child occupant protection, pedestrian protection and how the latest safety-assistance technologies, such as anti-lock brakes, really help drivers. Euro NCAP publishes safety reports on new cars, and awards star ratings based on the performance of the vehicles in a variety of crash tests, including front, side and pole impacts, and impacts with pedestrians. The top rating is five stars, and this rating is now a key promotional device used by manufacturers and recognized by consumers across Europe.

Product-safety legislation provides a set of stringent test criteria that products must pass before being put on sale. Such legislation has been harmonized throughout international trade blocs such as the European Union to ensure that less stringent requirements in one country cannot provide a back door for unsafe products to reach another, and that over-stringent requirements in some countries cannot be used to prevent the sale of perfectly safe products in others. In the European Union, for example, products that comply with the legislation are marked with the letters 'CE' (Conformité Européenne) – this logo indicates to enforcement authorities that the manufacturer claims compliance with the relevant local and international laws, and in many cases it is now illegal to sell products at home or abroad that are not CE-marked. In the USA, UL (Underwriters Laboratories Ltd) is one of the main certifying bodies and the FM Global and CSA (Canadian Standards Association) will also conduct testing, but these both have their own standards for testing products and the technique used depends on the type of equipment and the volume.

Particular types of products are also subject to type testing. These tests may be used by manufacturers during the design of their products but are principally aimed at providing an independent third party, such as the UK's FIRA (Furniture Industry Research Association), with a benchmark against which products can be assessed comprehensively and fairly. Consumers in the UK would recognize the BSI's (British Standards Institution) Kitemark and the Intertek's BEAB Approved Mark. In Europe the standard type test is the TÜV's GS Mark, while in the USA it is the UL Mark. These tests and the results they produce are recognized both nationally and internationally, and provide a vital safety net for consumers.

The definition of a 'product' has recently been expanded within the European Union to include digital manufacturing, raw materials, and software including AI systems, acknowledging the changing landscape of product design.

CIRCULAR DESIGN TESTING

The consumer product industry largely follows a linear 'take–make–use–waste' model: one in which we take raw finite materials, to make vast volumes of products, which we then use for a while, before inevitably discarding them as waste, with no further regard for their ecological or social impact.

Since the advent of mass production with the industrial revolution, the production and consumption of cheap, poorly made products has often quickly become a default standard, and their 'planned obsolescence' shortly after is the norm. Not only have recent industrial and technological developments made mass consumption possible, but they have also made it infinitely more desirable.

Today, it is becoming increasingly clear that there are ecological limits to this kind of linear growth, and that we need a design-led framework that offers systematic, holistic solutions to tackle the key environmental and social challenges of our time. A circular economy operates in accordance with three core principles:

Chapter 6 Testing

- Design out waste and pollution
- Keep products and materials in use for as long as possible at their highest value
- Regenerate natural systems

In a circular economy, these three principles are applied at every stage of the value chain: beginning with material cultivation and production, through product design, business and retail models, and finally to repair, reuse and recycle.

By nature, a circular system facilitates the continuous flow of materials, through both technical cycles (reuse, repair, remanufacture and recycle) and biological cycles (returning the nutrients from biodegradable materials safely to the earth in regenerative processes). Where, in a linear economy, raw materials might stand a chance of being recycled or downcycled at the end of their life, a circular economy crucially seeks to prevent this waste from occurring altogether. At the design phase, this looks like designing products for long-term use, and planning for multiple life cycles. Circular economy also offers radical new business models that focus on extending product life cycles through product-service systems, for example, rental, resale, maintenance and repair, and exchange, upcycling and high-value recycling.

While we see increasing risk associated with 'linear' business models, with the reliance on scarce and non-renewable resources and the failure to collaborate, innovate and adapt leading to price volatility, supply chain issues and legal repercussions due to changing legislation, the big-picture benefits of the circular economy are growing exponentially. Research indicates that a transition to circularity offers huge macro-economic, environmental, social and business benefits, including new job creation, enhanced economic growth and resilience, and increased customer loyalty and empowerment.

Testing products against circular design principles such as design for minimal waste, cyclability, durability and design using renewable, safe and recycled materials will help ensure your products meet the environmental and ethical standards that consumers and the wider world require to address the ecological and climate challenges we face.

Fig. 10
Design HOPES's circular-design reusable theatre caps.

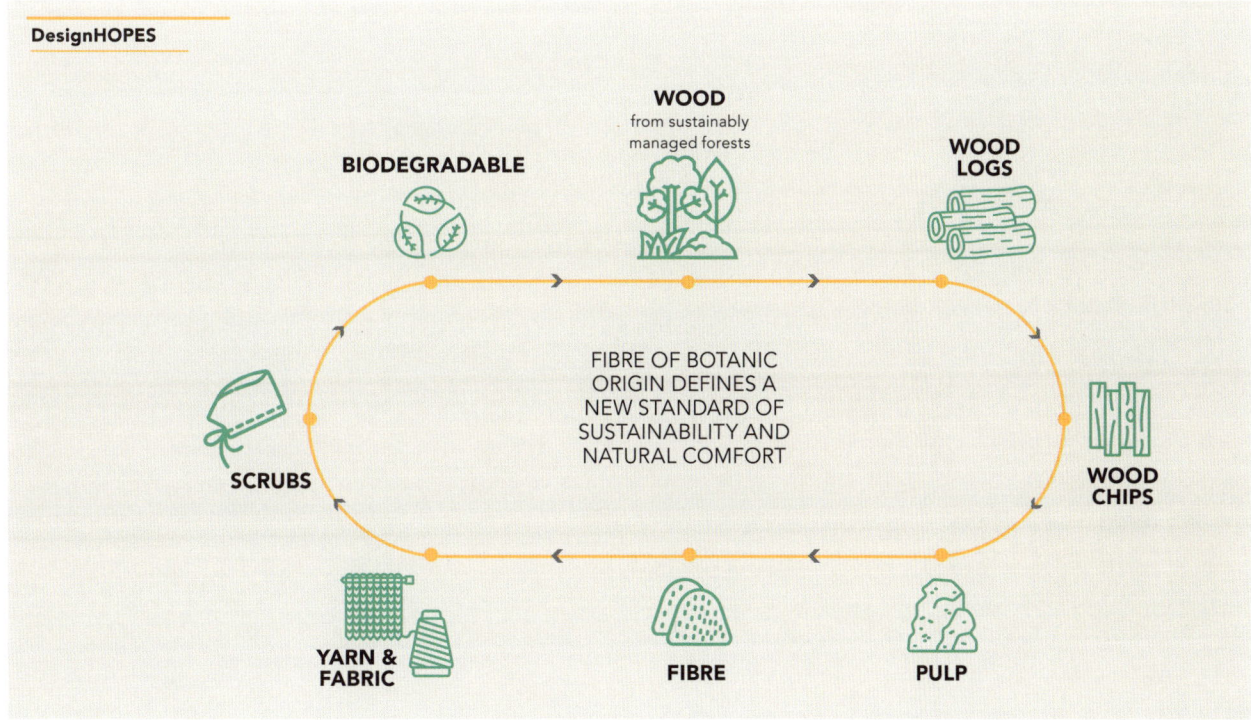

CASE STUDY

BERGHAUS FREEFLOW: REDEFINING BACKPACK COMFORT

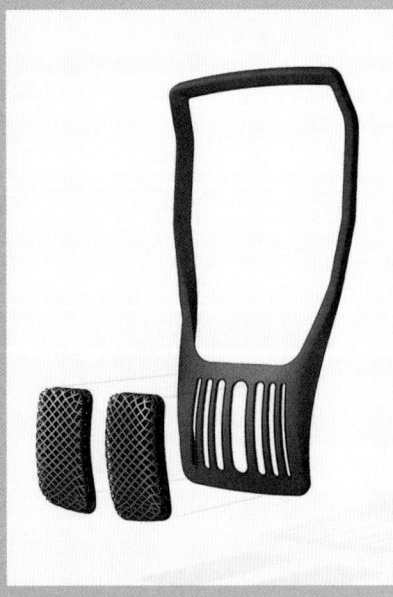

The updated backpack frame allowed for increased airflow without reducing internal volume or compromising structural stability, while new pads ensured a unisex fit.

Introduction
Berghaus was established in 1966 by climbers Peter Lockey and Gordon Davison, who were dissatisfied with the quality of outdoor gear available at the time. Their solution was to open a specialist mountaineering shop, which quickly evolved into a product-design hub. By manufacturing and selling their high-performance equipment, Berghaus built a reputation for innovation and user-centred design. Today, the company operates globally, with an in-house team of designers working across apparel, equipment and footwear.

Objective
Traditional hiking packs have long presented a challenge: they trap heat and inhibit sweat evaporation, causing discomfort and potential overheating during activity. Although many brands have attempted to address this issue through airflow back-systems, the market is now saturated with variations of the same approach, leading to consistent limitations in performance and user experience, with particularly notable drawbacks in user stability and packing space.

In this context, the Berghaus design team saw an opportunity to bring genuine innovation to a crowded and stagnant category: to design a hiking pack that maintains industry-leading breathability while significantly improving load stability and comfort, without compromising packing volume or fit versatility.

Methods
Berghaus approached the project with a rigorous testing and prototyping process, combining lab-based assessments, academic collaboration and real-world user testing. The design process began with primary research involving experienced hikers and outdoor users. Participants were observed during extended use, helping designers understand the relationship between back contact, heat retention and fatigue. Their feedback was critical in defining the performance requirements of the new system.

The design underwent five structured prototype stages, each focusing on refining a specific element such as airflow, load stability, lumbar support and pack fit. Each prototype was subjected to both lab-based analysis and field testing.

In Berghaus's internal testing facilities, thermal imaging and treadmill tests were used to evaluate heat distribution and breathability. Pressure-mapping tools were applied to assess weight distribution and pinpoint high-stress areas that might cause discomfort.

Berghaus partnered with a university research department to conduct climate chamber trials. These involved heat sensors and motion-capture systems to study the effects of heat buildup and how the pack interacted with natural human movement. Force-plate testing was also used to evaluate posture and gait changes caused by the backpack's weight and positioning.

Prototypes were field-tested by a group of experienced users across various terrains and weather conditions. Participants were fitted with thermal sensors to provide live data during testing hikes. Their qualitative feedback was used to further iterate the design, ensuring that real-world functionality matched lab performance.

Results
The final product achieved significant improvements in comfort, load stability and ventilation.
- A redesigned airflow system ensured consistent back

Above and below: Berghaus Freeflow backpack.

ventilation over prolonged use.
- The updated frame allowed for increased airflow without reducing internal volume or compromising structural stability.
- A unisex fit system, including two frame sizes and adjustable back lengths, accommodated a wide range of body types.
- The pack was released in multiple sizes and capacities, supporting use cases from single-day hikes to multi-day treks.

The redesigned Freeflow system reflects Berghaus's ongoing commitment to user-led innovation, combining advanced testing methodologies with cross-disciplinary collaboration. By addressing core ergonomic and thermal challenges, the Freeflow pack exemplifies how modern design processes can elevate product performance in demanding outdoor environments.
pentlandbrands.com

CASE STUDY

ENHANCING PAPR DEVICE USABILITY & DESIGN THROUGH TASK ANALYSIS

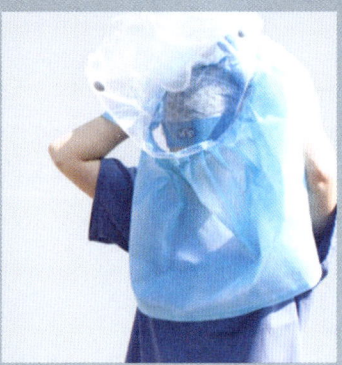

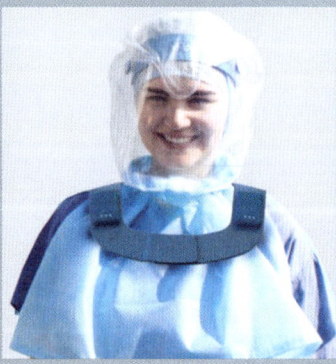

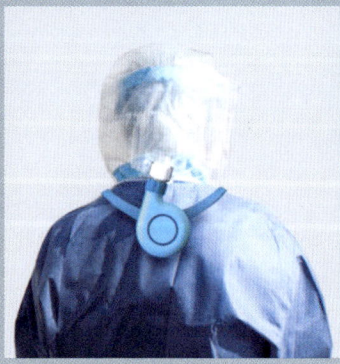

Above and opposite: Step-by-step usability testing for donning and doffing.

Introduction
This case study outlines how a structured task analysis improved the design of powered air-purifying respirator (PAPR) devices used in clinical settings. The approach was designed to capture every nuance of user interaction, ensuring that the final design proposal could meet the rigorous demands of high-stress clinical environments while remaining intuitive and efficient.

Objective
At the heart of this methodology is a mapping tool that is familiar to designers. The task analysis was organized using columns for each sub-task and swim lanes for various analysis headings, such as perception, cognition, action and insights. This format allowed the designer to visually trace each step of the device's use, from the moment it is first handled to the final stages of donning and doffing. The flexible nature of the document meant it could be tailored to the specific needs of the project and the designer, ensuring that every detail was captured in a manner that best supported further device development.

Methods
The designer began by identifying the diverse user groups and stakeholders who would interact with PAPR devices during clinical procedures. Recognizing that these devices were to be used in dynamic, high-pressure environments, the designer ensured that every relevant user, from clinicians who don and doff the equipment to support staff responsible for cleaning, stocking and preparing it, was considered.

A key component of the project was conducting contextual inquiries. The designer observed clinical staff performing their tasks while wearing the PAPR equipment. These observations were critical to understanding the impact of the devices on workflow, including potential delays in care delivery or disruptions in communication. In addition to real-time observation, the designer conducted interviews and discussions with users and stakeholders to gather deeper insights into the donning and doffing processes. Tools such as cognitive walkthroughs and role-playing exercises were used to simulate real-world scenarios, allowing the designer to understand users' thought processes and identify both strengths and weaknesses in the device designs.

The study focused on two specific PAPR models: the 3M Versaflo and the MaxAir. The designer systematically broke down the donning and doffing procedures into clear tasks, such as setup, attaching the respirator and affixing the hood, then further divided them into detailed sub-tasks. Each sub-task was described in terms of its objective and evaluated using the perception, cognition and action (PCA) framework. This approach examines what users perceive through their senses, how they process this sensory information through cognitive mechanisms and the resulting physical actions they perform. For instance, if a clinician pressed a button expecting the device to activate in some way but it failed to do so, the immediate cognitive reaction could disrupt the entire workflow.

To enhance the analysis, the designer incorporated environmental factors into the evaluation. Variables such as lighting, noise and overall visibility can significantly affect how a user interacts with a device. A poorly lit or noisy environment may amplify existing difficulties, possibly making the tasks of donning or

Chapter 6 Testing

doffing even more challenging. By evaluating these conditions alongside sensory and cognitive responses, the designer gained a comprehensive understanding of the interaction challenges.

One of the significant strengths of this approach is its transparency. The task analysis not only guided the design process but also served as a comprehensive research document. It can clearly reveal the origins of every design element and specification, making it an invaluable communication tool when collaborating with team members from diverse disciplines, such as engineering, marketing or clinical operations. This document provided clear evidence and rationale behind each usability decision. This level of clarity ensured that all stakeholders could understand, debate and support the proposed design improvements, thereby fostering interdisciplinary alignment and collaboration.

Results

Ultimately, this project did more than just map the ecosystem of users and stakeholders: it laid the groundwork for innovative design proposals. By combining contextual inquiries with detailed task analyses and evaluating every interaction through the PCA framework, the designer obtained a holistic understanding of both the challenges and the opportunities presented by the current PAPR devices. Environmental factors, such as lighting, noise and spatial constraints, were also taken into account, ensuring that the concept proposal would perform reliably under real-world conditions.

The task analysis revealed several critical pain points, from unclear user interfaces between the PAPR device and the sterile scrub donning to cumbersome attachment mechanisms, and even aspects relating to device weight and posture during prolonged use. These insights translated into specific design specifications that aimed to enhance safety, improve efficiency and ultimately result in a product better aligned with the needs of clinicians and their working environment.

This case study demonstrates that a well-constructed task analysis is an invaluable tool in device/product development. By systematically mapping user interactions and evaluating each step through detailed frameworks, designers can uncover latent issues and generate actionable insights. The result is a set of clear, evidence-based design specifications that improve the usability of the PAPR devices and also ensure that the final product supports the demanding workflow of clinical environments. This systematic approach not only supports the designer's rationale for usability improvements but also fosters interdisciplinary collaboration, ensuring that every aspect of the product's performance meets the rigorous demands of its intended use.

The project was developed by Maire Kane, under the supervision of Enda O'Dowd and Derek Valence, Joint Programme Leaders of the MSc Medical Device Design, NCAD.
ncad.ie

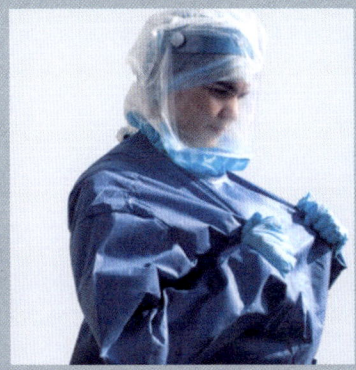
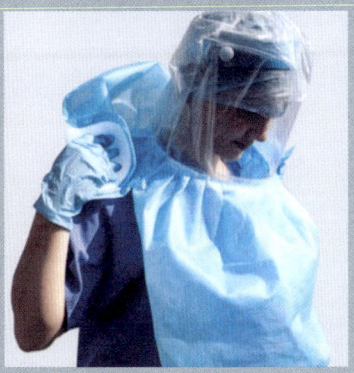
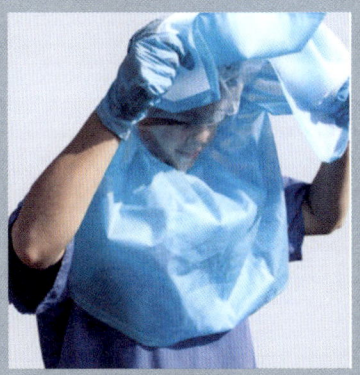
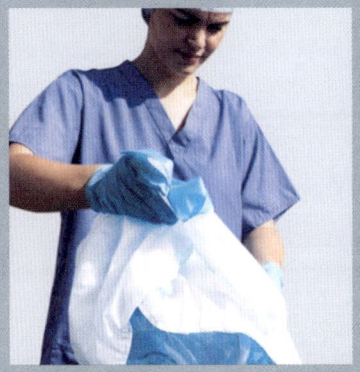

TUTORIAL

How to run a great user trial

In order to run a successful user trial you should take the following steps:

Set goals
When undertaking user trials you need to determine the key issues you wish to explore in order to provide useful design feedback. You may also need to set criteria, such as acceptable levels of performance and error rates, in order to determine whether the product under test meets usability metrics targets.

Design your test
You should decide on the number of users to be involved – the larger the sample, the more representative the test. Provided that detailed comparisons are not required, five test participants is usually sufficient. If the test involves more than one product or system, you must be clear whether the same user should be tested using both/all of them, or tested using only one. Testing participants with the same tasks on two different products or systems risks the results being affected by a potential transfer of experience between the two products under test. When evaluating the subjective preferences of users, it is best to let users test all the products.

Write a procedure plan
This describes how the test will be run from start to finish, and how it will be observed, filmed and recorded. It prescribes the sequence of tasks to be performed, when and what questions are asked, and how instructions are given. The procedure plan should also describe how the observations are to be made. You should pay strict attention to the logistics and scheduling of trials to avoid users having to sit and wait or being asked to come back another time.

Pilot the test
It is a good idea to run pilot trials with easily available subjects such as colleagues until you are sure that there will be no technical and procedural problems during the testing. It is important that the tasks are representative of the product under consideration. One approach is to try to select tasks that will be performed frequently by a typical user, although emphasis should also be placed on those tasks that are particularly important, such as safety.

Pre-test interviews and questionnaires
Before the actual testing, the users should be interviewed for relevant background information, such as age and gender. You should also find out if they have any prior experience with similar products.

Put your users at ease
You should allow sufficient time to introduce the users to the activities required. Try to ensure that the tests last, ideally, no longer than an hour to ensure users are unaffected by tiredness or other distractions. It is also extremely important that the testing and interviewing are performed in a relaxed and friendly atmosphere, where the user does not feel anxious. Avoid criticizing the user in any way for their performance, and emphasize that it is the product rather than the person that is being evaluated.

Post-test interviews
Once the test is complete you should

interview the user to capture invaluable feedback. During the interviews, the user's attention should be directed towards the product, and not towards their own shortcomings or failings in operating the product. You should also allow sufficient time to conclude the interview session, by answering any questions that participants may have.

Data analysis

A list of problems and issues is the usual output of a user trial, and these should be ranked in order of severity by the users. This might be done by asking users or experts to rate each problem on a ten-point scale. If this is not possible, you should rank the problems according to your own judgement.

Learning from the test

Once you have analyzed the data, you should determine how this can be used to inform your design processes and improve your product. Your video recording of the test and the problems identified is a particularly powerful communication tool, and can demonstrate more effectively than words to colleagues, clients and stakeholders the practical problems experienced by end-users when operating equipment.

Trialling a hand-held game. It is important when conducting a user trial that you gain feedback on durability and practicality as well as aesthetic qualities.

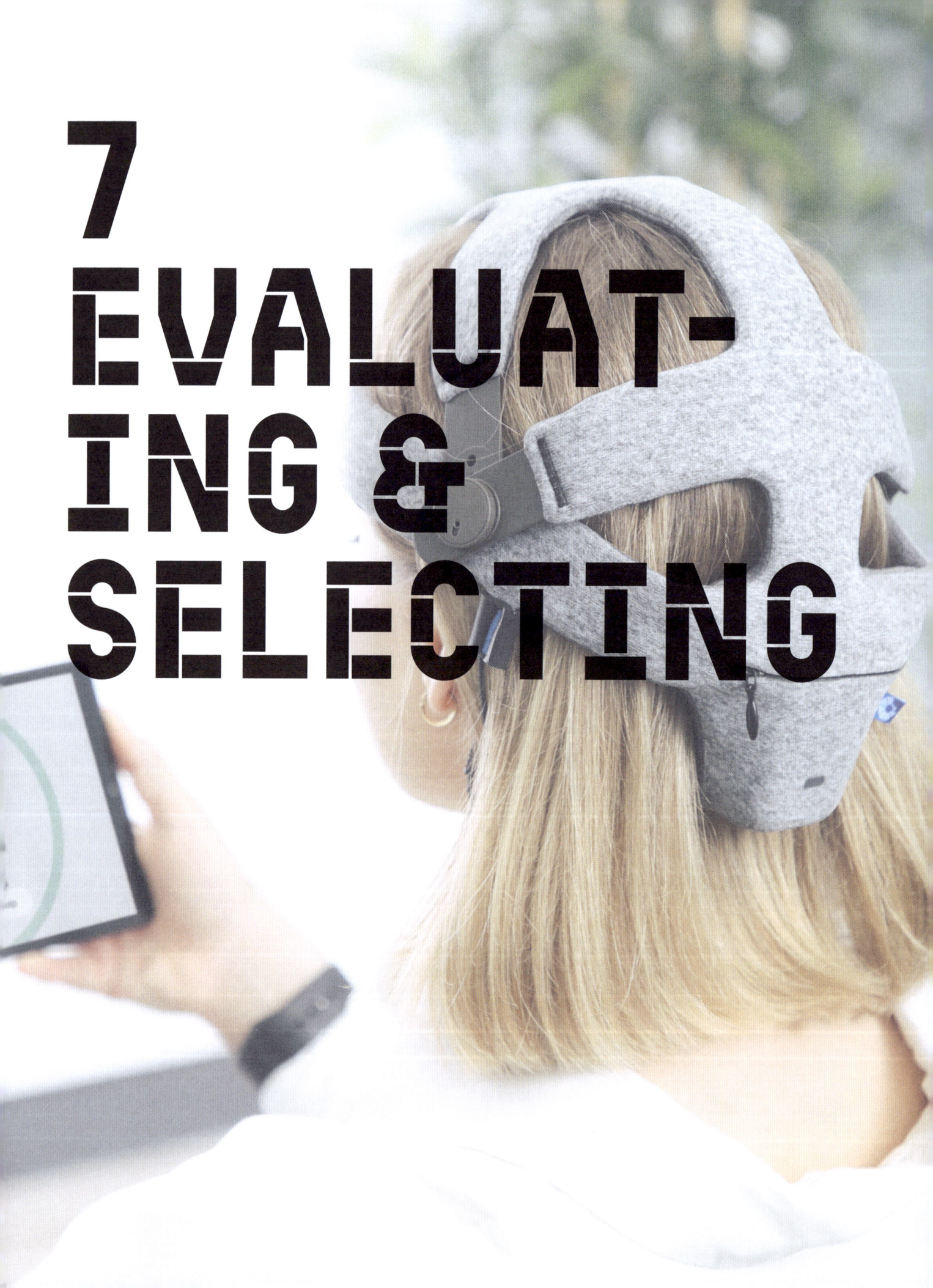

7 EVALUATING & SELECTING

CHOOSING THE RIGHT METHODS

Designers need reliable, rigorous and robust methods of evaluating and selecting their design proposals. Picking the wrong product design proposal to develop can be very costly to manufacturers and stakeholders in terms of time, money and valuable resources, so there is a major incentive for designers to get it right the first time. Evaluating and selecting concept design proposals is a convergent process, and it is frequently highly iterative. Choosing the right methods is, therefore, extremely important.

Developing and launching new products carries risks: will it appeal to the target market? Does it have the right features? Can it be manufactured profitably? Is the development cost justified? Furthermore, designers must consider the environmental impact of their products, ensuring that sustainable materials and processes are prioritized, and that a product's lifecycle is managed responsibly. It is no longer safe for product designers to react to risks as they develop; instead, designers must actively identify and measure these risks before they become a problem. Incorporating EDI principles into product design ensures that the needs of diverse user groups are addressed and that accessibility and inclusivity are prioritized.

To effectively track and manage the risks associated with delivering innovation, design companies, no matter how small, should develop a robust risk-management strategy that encompasses environmental and ethical dimensions; otherwise they could face potentially disastrous consequences. Designers are constantly evaluating which directions to take while, at the same time, generating many concepts to choose from. The systematic evaluation techniques presented in this chapter should help maximize your chances of selecting the right design proposal while fostering sustainable and ethical practices. When considering a proposal, it is helpful to draw up a product design specification (PDS), which is essential during the design process and also clearly sets out the parameters that must be met so the designer can make an informed decision before taking on a project. Criteria for environmental sustainability and ethical sourcing, manufacture, distribution and use are critical components of the PDS. Below are some of the key points that should appear in a PDS.

Fig. 1
The PDS is a dynamic document that can, and often does, change over the course of the design process. The PDS should layout the main considerations of a brief at the beginning of a project.

Fig. 2
A typical product checklist. The use of such lists helps designers ensure they meet the challenge of addressing all of the issues raised in the product design specification (PDS).

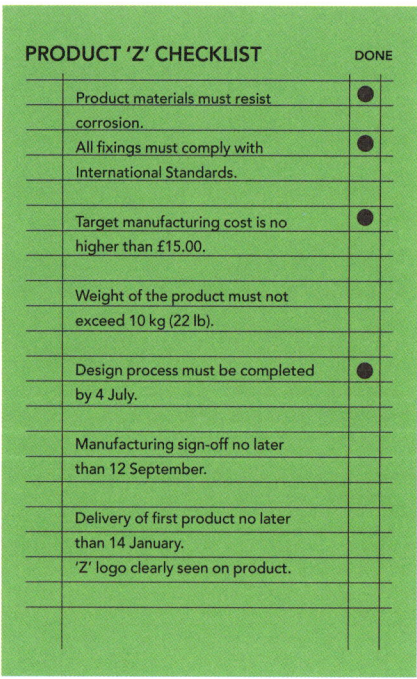

CHECKLISTS

A checklist is a very useful device, particularly when it comes to alleviating or eliminating any errors that might arise in the design and development of products, services, systems or environments. Sometimes referred to as 'to do' lists, checklists make tasks or objectives that need to be met explicit. This is especially useful when designing products and services that have a high degree of interaction with the end-user. Checklists can also help ensure consistency and completeness in carrying out specific tasks and are commonly found in a number of areas, such as the aerospace industry, healthcare, engineering, and in projects with potentially significant public liability issues.

Generally speaking, the product design specification (PDS) can be viewed as a form of checklist. In product design projects, a PDS is created to ensure that the product designer produces a design solution that reflects a true understanding of the actual problem and the needs of the user. In short, the PDS is a series of checklists split into smaller categories to make it easier to consider. These might cover categories such as cost, safety, size, packaging and quantity, and include specific requirements such as:

— The product's cost of manufacture must be no higher than £3.00 per unit.
— The length of the product should be no greater than 75 mm or 3 inches.
— The weight of the product must be kept to a minimum and no greater than 1.5 kg.
— The product must be waterproof.

A checklist is not a set of instructions – its aim is to help identify and address common problems and errors during specific stages of the design and development of a new product. For this reason, it should be simple, measurable and translatable. A checklist is also a good tool for managing actions and communications between members of the design and development team. Generally speaking, the checklist is meant to be read aloud as actions are being performed: 'The product's cost of manufacture is less than £3.00 per unit' – 'CHECK!', and so on. A checklist should be seen as a dynamic document that evolves over the course of the design process as elements are tested, re-tested and refined.

Fig. 3
Designers seek to verify their concepts through discussion with stakeholders, evaluating their concepts' respective merits, and refining their designs through invaluable user input.

EXTERNAL DECISION MAKING

Designers are increasingly looking to engage consumers and stakeholders in the design process, and involve customers and clients in the decision-making process. When designing 'with' rather than 'for' people, there is a spectrum of engagement that a design team can operate within:

— Stakeholder consultation is the minimum level of engagement generally acceptable in practice today, with users informing and influencing the decision making through methods such as focus groups, surveys and questionnaires.
— Participatory design actively involves people at all stages of the design process to ensure their concerns are understood and considered, and to give them some influence on and ownership of design decisions. This level of engagement requires a greater level of dialogue with the stakeholders, and often takes the form of inclusive design workshops, where participants can actually shape the design process.
— Collaborative design brings people into an active design partnership and the designers and stakeholders agree a full and frank sharing of resources and decision making. Examples of this approach include companies setting up advisory panels and strategic partnerships with a range of stakeholders.
— Design delegation is the most extreme intensity of engagement, where a manufacturer delegates the design decision-making process to consumers, through the use of methods such as ballots and referendums. This apparently democratic model relies on the questions being asked of the 'voters' being carefully written and designed to ensure clarity of purpose, and an explicit description of how any decisions will be subsequently implemented.

Involving large numbers of users in the evaluation and selection of designs undoubtedly provides invaluable feedback and insights to the design team, but this can sometimes be impossible due to time or financial constraints. Therefore, it is common practice to try to gain useful feedback from a number of diverse representative users and stakeholders to reduce biases in the sample users' responses and observations.

The following list can help to structure your use of external decision making:
— Broad user mix – this involves users from a range of market segments, which can help understand general user requirements.

— Boundary users – this involves users on the limit of being able to use the product, which can help identify opportunities for design improvement.
— Extreme users – these can inspire creativity during concept development.
— Mixed-experience users – those with different levels of experience with similar products can help understand the impact of experience on use.
— Community groups – again, those sharing experience of interacting with similar products can provide a broad understanding of product use.

INTUITION

A product designer's intuition can be an important factor in the evaluation and selection of concept design proposals. A decision based on intuition alone is where a concept is chosen for its 'feel', with the designers relying on their own tacit knowledge rather than explicit criteria. When product designers are asked to articulate their skills and explain how they work and make decisions, 'personal intuition' is one of the most frequently given responses.

Design intuition is an important and effective part of a designer's make-up. It takes years of experience and practice to hone, and depends on careful observation of people, the products they use and how they use them. A designer might not always be able to articulate their judgements and decision making analytically, but this does not necessarily indicate that their intuitions are weak. In practice, product design projects proceed by a combination of intuition and reasoned analysis.

Intuition is a difficult concept to define but one that most designers recognize as crucial. It is not strictly speaking a methodology, although when matched with play, it can influence your creativity, and shape your personal design methodology. Play is an unavoidable and essential element in the design process, but one which is largely ignored. The dry, reductivist view of design that seeks to promote the designer as an objective, emotionless entity struggles when looking for explanations of recent design trends, such as the playful designs of Alessi and Droog Design's 'Do' collection. The psychologist Carl Jung tells us: 'The creation of something new is not accomplished by the intellect but by the play instinct acting from inner necessity. The creative mind plays with the objects it loves.' This argument seems fairly natural and familiar to most practising designers, although the intuitive and play element of their work may be 'underplayed' in their professional lives.

There are several kinds of intuition – the intuition of the product designer is unlike that of the engineer, and both will be dissimilar to the marketing specialist. The role and nature of intuition in design has been somewhat overlooked by many design writers and critics, perhaps because it is one of many unobservable mental entities that scientists are yet to completely understand. For designers, intuition relies heavily on insights that are based on both knowledge and experience.

The key conditions necessary for intuition appear to be:

— Sufficient conceptual knowledge and experience of the field.
— A strong motivation to resolve the issues associated with the project.
— A period of relaxation, when the designer is not consciously thinking about the project. This might include listening to music, falling asleep or awakening.

To paraphrase Thomas Edison's famous quote – great inventions are 1 per cent intuition (inspiration) and 99 per cent perspiration. Intuition in design can often lead to brilliant ideas quickly. On the other hand, it can be extremely unreliable in terms of both time and usefulness. The main problem with relying on intuition alone is that it might never happen.

Fig. 4
(Opposite) Designer Max Lamb takes a very hands-on and intuitive approach to design. His pieces have materials and process at their heart, and achieve a rawness and sense of craftsmanship that machined products just don't have. Seen here is the process of creating his Urushi Stool using primitive green-woodworking techniques.

Chapter 7 Evaluating & Selecting

CROWDSOURCING

The term crowdsourcing was first coined by Jeff Howe in a June 2006 *Wired* magazine article, where he stated that new technologies and communication platforms such as the internet enabled companies to take advantage of the expertise of the public, and involve them in the creative design and decision-making process. Howe stated that 'It's not outsourcing; it's crowdsourcing.'

Crowdsourcing is essentially the act of outsourcing tasks traditionally performed by experts to a group of people through an online open call. The idea is that actively engaging people outside the usual corporate environment can access a far larger pool of expertise, who can perform tasks, solve complex problems and contribute fresh ideas. Companies are eager to find new insights, ideas for new products and qualitative inspiration from their users, and a dedicated community crowdsourcing platform can be the perfect tool with which to start the conversation with their target group.

Crowdsourcing can take a variety of forms, but the usual format is for a company to present a new product concept to their online community. This community of users, experts and interested parties is then able to freely modify the current design or upload their own ideas, using any combination of comments, sketches, pictures, mood boards, movies, prototypes or total redesigns. Rewards for inputs can range from a payment or entry into a competition, through to a formal royalties deal for a chosen design.

The ideas and comments generated online are fed back to the company's research and development team, who then use this data to inform their decision making. The hoped-for result is the co-creation of products that are better aligned with what their consumers really want.

Fig. 5
Companies are increasingly engaging the general public in the development of their products through methods such as crowdsourcing, where large groups' opinions are gathered or they are asked to trial a particular topic.

A recent development has been the introduction of crowdvoting, which makes use of the power of popular opinion to rank and sort all kinds of data. Often this interaction takes place in real time, with companies promoting their brand or product through live content. By asking large numbers of potential consumers to vote for a particular concept over another, companies are able to generate significant levels of consumer interest and buy-in for the winning concept that enters production. This form of crowdsourcing is increasingly used as a marketing tool, alongside other forms of social media, such as Facebook and Twitter, to develop and promote new products.

Recent examples of the use of social media to develop or promote products include Vitamin Water and Fiorelli. Vitamin Water asked users on their Facebook page to help them choose the next flavour of the drink to be released and effectively crowdsourced their creative design process. The campaign was endorsed and supported by leading celebrities and created a considerable online trend on X. Luxury accessories brand Fiorelli recently launched a presence on Facebook that included a new social shopping experience for customers and fans. Buying an expensive Fiorelli bag is a major investment for most customers. The brand has taken this into consideration by introducing a Facebook app that allows people to superimpose a photo of their bag of choice onto their profile photo and receive their friends' comments before they decide to purchase, thus demonstrating the use of social media to help influence customers' actual purchasing journey.

PRODUCT CHAMPIONS

Many people would argue that a product champion is absolutely crucial for any company designing and developing new products today. A product champion is an influential member of the new product design and development team. She or he supports the instigation and development of concept design proposals that, in turn, help determine the overall design direction. The importance of product champions to the success of developing innovative products was highlighted in business and management research in the 1960s, and decades of studies since have supported these claims. A product champion is seen as a vital cog in the innovation process – needed to overcome organizational barriers and resistance within companies. Several proponents believe that new product design ideas either find a champion or die.

The role of a product champion, essentially, is to actively and enthusiastically promote and sell a project in order to obtain vital and valuable organizational resources and support. Moreover, product champions are individuals who might occasionally have to take personal risks by advocating and demonstrating the feasibility of a project to reluctant top management.

Fig. 6
Brands regularly employ product champions to evaluate and promote their products, and an endorsement from a leading sports star such as Michael Jordan or Lionel Messi can guarantee commercial success.

The product champion acts as an important go-between or linkage point between various departments within an organization. He or she will possess good knowledge of the organization's capabilities and resources and will know which individuals should be concerned with the innovation, thus connecting the organization's sponsor(s) with the internal and perhaps external experts. One of the product champion's key strengths will be the ability to translate the technical language of an innovation into terminology that is commonly used in the organization – for example, they will ensure management and business leaders within the company are able to understand the work of the design and engineering team. By becoming a salesman of a new design idea, the product champion is able to develop a plan of action, using their diplomatic talents to provide access to different people within the organization.

MATRIX EVALUATION

Matrix evaluation, sometimes referred to as 'Pugh's Method' after the British engineering design professor Stuart Pugh, is a quantitative technique used by designers to evaluate their concept design proposals by ranking them against the set criteria stated in the product design specification (PDS) and/or against other concept design proposals. While many stages involved in the product design and development process benefit from unrestrained creativity and divergent thinking, the selection of concept design proposals is the process of narrowing down a number of alternative proposals to select one for further development and refinement.

This is a convergent process, but it is frequently iterative and may not produce a strong or dominant concept proposal immediately. The matrix evaluation technique may have to be run several times before a strong concept design proposal emerges as ripe for further development. Thus, selection and evaluation are iterative processes that must be embedded in the development of new products.

Decision matrix model (Pugh's Method)

— Select decision criteria
— Formulate decision matrix
— Clarify design concepts being evaluated
— Choose datum or best initial concept
— Compare other concepts to datum based on a +, –, S scale
— Evaluate the ratings; discuss concept's strengths and weaknesses
— Select a new datum concept and re-run analysis
— Plan further work
— Second working session to repeat above steps and select a concept

Designers constantly evaluate and select which direction to take while generating design proposals. Typically, this will mean that several concept design proposals need to be chosen from. Usually, with matrix evaluation, a large number of proposals will be rapidly narrowed down to a more concise and focused number. This might involve combining specific features of one concept with another totally different concept to improve and temporarily enlarge the set of concepts being evaluated and selected. Eventually, after working through several iterations, a dominant concept proposal will emerge.

When selecting which concept design proposals best satisfy the PDS, it is essential to remember the need to generate wholly new concepts, adapt existing ones or undertake further research to proceed. Evaluation and selection should be a narrowing process, weeding out unsuitable ideas, rather than trying to pick the 'best idea'. Referring back to the PDS and placing yourself in the user's position can help avoid selection on a purely subjective basis.

Once an appropriate number of design concepts have been generated through sketching and modelling, the design team can refer back to the PDS and evaluate and select which concepts fulfil the criteria laid out in the original specification. To avoid subjectivity and personal intuition creeping into the decision-making process, it is imperative that all members of the product design and development team perform this vital part of the process. If possible, input should also be included from the client and/or stakeholders, helping to evaluate and select the concept design proposals outlined from a number of perspectives.

Matrix evaluation can help designers, engineers, manufacturers, marketing staff, users, clients and buyers to reduce ambiguity and confusion in the evaluation and selection process, enjoy better communication, and deliver successful new products to market more rapidly.

Fig. 7
Evaluation matrices are used to evaluate a number of design options against prioritized criteria. This process is relatively simple to apply and aids the design team in making objective decisions.

PRODUCT 'Z' CHECKLIST	CONCEPTS					
	▲	⬟	●	○	□	
Product materials must resist corrosion	+	-	+	=	=	
All fixings must comply with International Standards	+	=	-	+	=	D
Target manufacturing cost is no higher than £15.00	+	+	=	=	-	A
Weight of the product must not exceed 10 kg (22 lb)	=	=	-	+	+	T
Design process must be completed by 4 July	+	+	+	=	=	U
Manufacturing sign-off no later than 12 September	=	=	=	-	-	M
Delivery of first product no later than 14 January	=	-	=	=	-	
'Z' logo clearly seen on the product	=	-	=	=	-	
TOTAL	4+ 0- 4=	2+ 3- 3=	2+ 2- 4=	2+ 1- 5=	1+ 4- 3=	

CASE STUDY

LOTUS THEORY 1

Introduction
This case study demonstrates how manufacturers develop and showcase concepts to evaluate new design directions.

In 2024, Lotus introduced Theory 1, a concept car that sought to encapsulate the brand's future direction as an intelligent performance car manufacturer. A three-seater electric-powered sports car with a central driving position and exposed aerodynamic structures, Theory 1 blends the iconic Lotus Esprit's 'wedge' styling from the 1970s with contemporary Formula 1-inspired engineering. It serves as a manifesto for Lotus's evolving design and innovation strategy, guided by three core design principles: digital, natural and analogue (DNA).

Objective
The project was driven by three key objectives:

- Embody brand ethos: reinforce Lotus's rebellious, pioneering and sophisticated identity as it transitions into a global technology-led performance brand.
- Establish a coherent design theory: unify interior, exterior, materials, UX- and UI-design through a clear, translatable design framework.
- Showcase innovation through partnerships: test outputs from Lotus's R&D activities and collaborate with strategic partners to push boundaries in materials, ergonomics and digital technology.

Methods
The creation of Theory 1 was structured around three interwoven research streams: brand legacy analysis, development of a unified design theory, and a multidisciplinary innovation programme.

1. Brand legacy study
The team conducted an in-depth analysis of Lotus's design and engineering history, from exploring archival materials to referencing the biography of founder Colin Chapman. Key influences were identified, such as the need for a strong design language anchored in the idea of 'a car you wear' and Chapman's motto 'simplify and add lightness'. This research not only influenced visual elements of the concept but also shaped the overall design intent of the project, reinforcing continuity between past and future Lotus models.

2. Defining the design theory (DNA)
To unify the creative direction, teams from interior, exterior, materials, UX and UI collaborated in joint workshops. Each discipline explained their interpretations and inspirations, which coalesced around the 'DNA' framework:

Digital: immersive, intuitive interaction and advanced computational intelligence.
Natural: emotionally engaging, human-centred design.
Analogue: mechanical clarity and visible performance engineering.

Below: Clay modelling is used to develop the design language initially explored in the exterior design sketches.

Chapter 7 Evaluating & Selecting

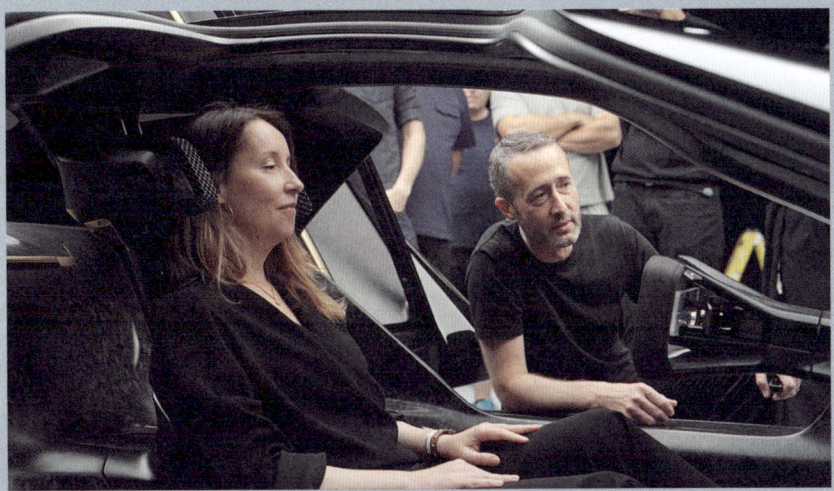

Top: Interior Design UX/UI testing.
Middle: Finished interior design.
Bottom: Lotus Theory 1 and original Lotus Esprit.

3. Development of a collaborative research programme

The Theory Lab facilitated external collaborations that introduced cutting-edge innovations, aligned to the DNA principles. These included the adoption of human-centred design methods to create an interior inspired by elite athletic apparel, ergonomics and wearable technology. The use of sustainable, lightweight and recyclable materials, along with additive manufacturing and robotic textiles demonstrated a commitment to material innovation. They also tested immersive technologies, such as advanced binaural audio and visual system facilitating, to enhance emotional connection between driver and car.

Partners in the project included KEF (audio), Carbon (3D printing), Pirelli, Kyocera, MotorSkins, and Archivio Storico Olivetti. These collaborations explored intersections between automotive, tech, fashion and art to test future design directions and technologies.

Results

The project culminated in several breakthrough features, including LOTUSWEAR™, which is a tactile interface system using robotic textile seating, haptic feedback via the steering wheel and adaptive materials to personalize the driving experience. Staying true to Lotus's lightweight philosophy, the concept adopted a material reduction strategy, using only ten materials instead of the hundreds usually employed in the manufacture of a car, emphasizing Lotus's commitment to sustainability.

Theory 1 stands as a high-performance electric concept car that combines advanced research, brand heritage and forward-thinking design methods into a single cohesive vision for the future of Lotus. It was awarded the prestigious 2025 Car Design Award for best Concept Car, with the jury praising how faithful the product was to Lotus's ethos of innovation.

lotuscars.com

CASE STUDY

PA CONSULTING & CUMULUS'S BRAIN HEALTH MONITORING PLATFORM

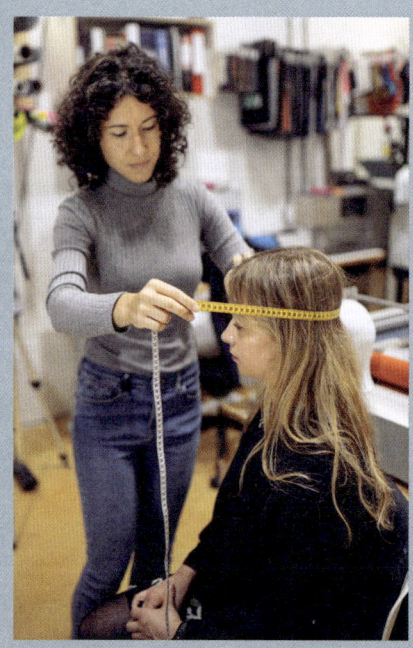

Top: Traditional craft skills from the fashion industry, such as pattern-making, are merged with advanced techniques, such as flexible electronic development.
Above: A deep understanding of anthropometrics and human factors engineering was fundamental to the success of the product.

Introduction
Neurodegenerative and neuropsychiatric diseases like Alzheimer's, dementia and depression are a growing global challenge, where drug discovery and disease management are hindered by late diagnosis and monitoring difficulty. In response, Cumulus partnered with PA Consulting to develop a medical-grade EEG-monitoring (recording of brain activity) platform for at-home use in decentralized clinical studies.

Objective
Lab-based EEG systems, which require a trained technician to perform EEG recordings, have many heavy wires and cables in a cumbersome form to maximize their resolution. Cumulus aimed to achieve lab-based resolution in a format that can be brought into the home and worn comfortably for extended periods, providing a revolutionary combination of accuracy and accessibility.

Methods
This project integrates novel processes and methods specifically designed for a smart garment or wearable device. Leveraging PA's experience in wearable design, the team identified a key challenge: ensuring high-quality, clinical-grade data collection while maximizing user adoption and adherence.

Unlike traditional industrial design, wearables require a personalized approach – one size does not fit all. As this was a head-worn device, the team conducted extensive research on head sizes and shapes, using 3D scanning and direct measurements. To ensure precise electrode contact, they combined real-world testing with an extensive database of 3D head scans. Through iterative prototyping, a sizing strategy with an innovative adjustability system was developed that fits up to 98% of head shapes and sizes.

Understanding cultural perceptions of head-worn devices was also essential. Contextual inquiries explored user comfort, acceptance and usability concerns, focusing on interaction design with the platform's tablet interface and consistent electrode positioning when users put on and took off the device. Product analysis of similar devices, task analysis and human factor studies also informed the design approach.

To accelerate innovation, the team built a series of 'Monster Prototypes': quick, unrefined mock-ups that tested key experience elements. These prototypes were developed using both fashion- and industrial-design techniques, such as fabric selection, pattern cutting, sewing, tape-bonding and embroidery, alongside hand-modelling, rapid prototyping, electronics and additive manufacturing tools.

The team came up with several creative solutions for such issues as how to manage the unit's complex wiring. The solutions ensured that the headset met the criteria for larger-scale production, significantly reduced the cost of making it and offered a neater aesthetic.

By exploring various combinations of materials, profiles and connection methods, the team designed a unique electrode solution. This process achieved robust contact across all sensors ensuring superior accuracy of data. Proprietary AI-powered analytics then automatically evaluated the rich multidimensional data to rapidly generate meaningful insights for clinical trials.

Sustainability was a core design principle. Electrical components could be removed from textiles, allowing for disinfection between clinical-trial

Chapter 7 Evaluating & Selecting

participants and reducing waste by reusing high-cost components. The goal was to create an effortless, inviting solution that encouraged long-term engagement. Inspired by the garment industry rather than the medical or tech sectors, the wearable was designed to conceal its advanced technology, enhancing user confidence. Soft fabric cushioning improved comfort, while selected textiles and foams provided stretch for fit, foldability for transport and shape retention for wearability. Water-resistant fabric finishes further improved cleanliness and durability.

By balancing technological and material innovation with care-centric design, this project successfully merged clinical precision with everyday wearability.

Results
To tackle the challenges of monitoring brain health, Cumulus and PA partnered to develop the world's most advanced at-home, medical-grade EEG platform for flexible data collection and analysis. Through accelerating CNS (central nervous system) clinical trials, Cumulus helps bring life-changing therapies to patients faster and more cost effectively.
paconsulting.com
cumulusneuro.com

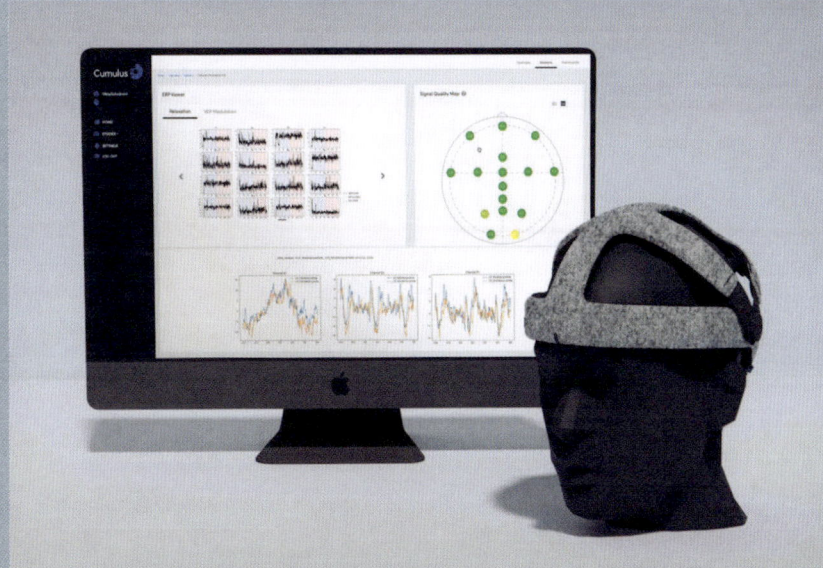

Wearable technology enables rapid data collection and analysis to bring about life-changing therapies.

How to conduct a matrix evaluation

Matrix evaluation is a reliable tool for structuring the evaluation and selection of concept design proposals. To evaluate concepts effectively, an agreed set of criteria – perhaps a checklist or PDS (see the next tutorial) – is required. Matrix evaluation is best carried out with a small group, although it can also be conducted as an individual activity.

The 14-step matrix evaluation procedure is as follows:

1. All of the concept design proposals must be generated to fulfil the same criteria set out in the PDS or checklist.

2. All of the concept design proposals must be represented to the same level of detail, be that in sketch or 3D model form.

3. Create a concept evaluation and selection matrix that will facilitate comparison of the generated concept design proposals against the criteria and one another.

4. It is important that the matrix includes all of the concept proposal visuals (e.g. sketches, models) so that the team can observe the pattern of evaluation that emerges. Textual descriptions of the concept proposals may be added to increase clarity.

5. The criteria that the concept design proposals are to be evaluated against must be defined before the generation of solutions begins. The criteria must be stated in unambiguous terms and understood and accepted by all members of the design team involved in the evaluation and selection process.

6. Select one concept design proposal as the datum point against which the rest of the proposals will be compared. Usually this is the concept proposal that the team thinks is the 'best'.

7. Each concept/criteria pair must be compared against the datum proposal and the following symbols should be used:
   + (plus sign): means better than, less prone to, easier than, etc. the datum.
   – (minus sign): means worse than, more expensive then, more prone to, more difficult than, etc. the datum.
   = (equal sign): means the same as the datum.

8. After selecting the datum, make an initial comparison of the concept design proposals using the +, – and = symbols. A score pattern will be created for each proposal relative to the datum. Bear in mind that the numbers, at this stage, are for guidance only.

9. Assess the individual concept design proposal scores, paying particular attention to each proposal's strengths and weaknesses.

10. Next, assess the negatives of the strong concepts and see if they can be easily reversed into positives. At this stage, you might wish to expand the matrix by introducing modified concepts.

Chapter 7 Evaluating & Selecting

		1	2	3	4	5	6	7	8
Withstand vandalism			+	=	+	+	=	+	+
Not heavier than 2.5 kg			−	−	−	−	−	−	+
Comply with the Home Safety Act (UK)			−	+	+	+	+	+	+
Cost no more than £25.00			−	+	−	−	+	−	=
100% recyclable			=	+	−	−	+	−	−
Lasts 3 years minimum			−	+	=	−	+	+	−
+ number			1	4	2	2	4	3	3
= number			4	1	3	4	1	3	2
− number			1	1	1	0	1	0	1

11. Similarly, explore the weak concepts now and see if their negatives can be improved upon (relative to the datum concept). If they can be improved upon then include them and expand the matrix.

12. After completing steps 10 and 11, the weak concepts should be eliminated from the matrix. This will reduce the matrix size.

13. If a number of strong concept design proposals do not emerge after conducting steps 10, 11 and 12 then it usually means one of two things:
— The criteria are ambiguous and are being interpreted differently by the design team, leading to confusion.
— Some of the concept proposals are either too similar in nature or even the same thing.

14. Once a strong concept proposal emerges, repeat the matrix evaluation using the strong concept as the datum. If the results are repeated it confirms the strength of the concept. If not, then repeat steps 10 and 11 until strong concepts emerge.

Having completed the 14 steps of the matrix evaluation listed above, the design team will have acquired:
— A sound insight into the requirements of the PDS or checklist.
— A better understanding of the problem or issues.
— A better insight into potential design solutions.
— A good knowledge of the strengths and weaknesses of the concept design proposals made.
— Greater motivation to produce other concept design proposals.

The above diagram shows a matrix evaluation using sample criteria.

TUTORIAL

How to write a checklist (PDS)

As stated earlier in this chapter, the product design specification (PDS) can be seen as a form of checklist. Typically, this is created as a written document that can be changed over the course of the design process (and usually is). Usually, the design of the product, service or environment will follow the statements laid out in the PDS. However, if the emerging design departs from the PDS for some good reason, then the PDS can be revised to accommodate the change. The important thing is to keep the PDS and the product being designed in correspondence throughout the design process. In this way, the PDS ends up specifying not just the design, but the product itself.

A comprehensive PDS will comprise a sequence of anything up to 32 categories that cover aspects of the intended product's performance, its cost, how it will be made, what it must look like and how it will be disposed of after its life in service. You should start writing your PDS by listing the 32 headings shown below.

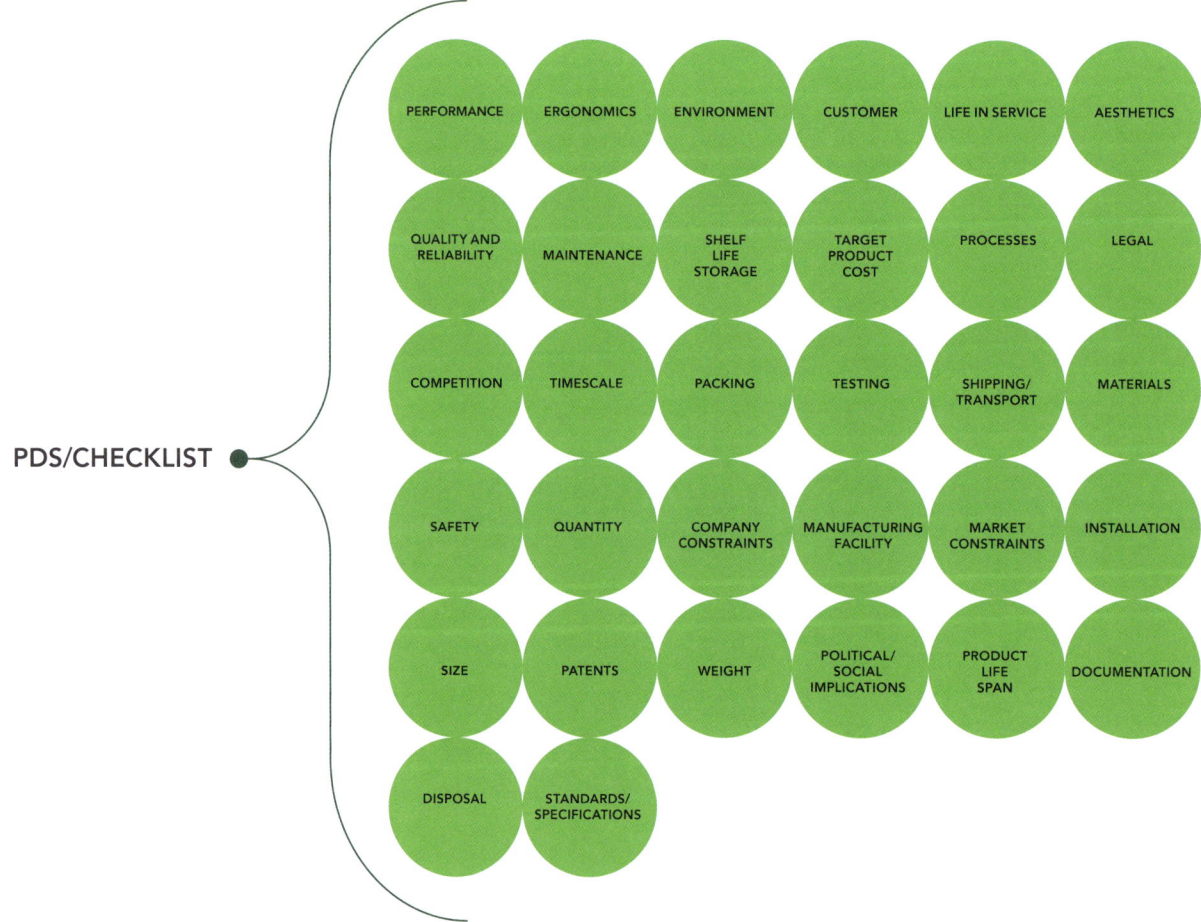

PDS/CHECKLIST

- PERFORMANCE
- ERGONOMICS
- ENVIRONMENT
- CUSTOMER
- LIFE IN SERVICE
- AESTHETICS
- QUALITY AND RELIABILITY
- MAINTENANCE
- SHELF LIFE STORAGE
- TARGET PRODUCT COST
- PROCESSES
- LEGAL
- COMPETITION
- TIMESCALE
- PACKING
- TESTING
- SHIPPING/TRANSPORT
- MATERIALS
- SAFETY
- QUANTITY
- COMPANY CONSTRAINTS
- MANUFACTURING FACILITY
- MARKET CONSTRAINTS
- INSTALLATION
- SIZE
- PATENTS
- WEIGHT
- POLITICAL/SOCIAL IMPLICATIONS
- PRODUCT LIFE SPAN
- DOCUMENTATION
- DISPOSAL
- STANDARDS/SPECIFICATIONS

Chapter 7 Evaluating & Selecting

Next, you need to write the individual checklist elements under each of the 32 headings, leaving out only those that clearly do not apply. Some of the elements in the PDS/checklist will overlap, but do not be tempted to skip any of them. In some rare projects, however, it may be appropriate to leave out a number of elements as they may not be relevant, but this needs to be agreed by all the design project stakeholders in advance.

Every single PDS element should be written with a metric and a value. For example, 'The length (time) in service of product A' is a metric and 'No less than five years' is the value of this metric. Values should always be labelled with an appropriate unit (e.g. seconds, metres, kilograms). Both the metrics and the values form the basis of the PDS/checklist. The headings and elements shown here give an example of how you should complete your own checklist.

It should be emphasized, however, that the PDS/checklist spells out in precise, measurable detail what the product has to do but not how it should be done. Each PDS/checklist element should be clear and succinct, using short, sharp precise statements under each heading, with metrics and values against each one.

PDS/CHECKLIST

HEADINGS	ELEMENTS
ENVIRONMENT	The product should be able to withstand vandalism.
WEIGHT	The weight of the product should be no greater than 2.5 kg.
SAFETY	The product must comply with all relevant parts of BS 3456 and the Home Safety Act (UK).
TARGET PRODUCT COST	The retail cost of the product must be no higher than £25.00.
INSTALLATION	The product must be ready assembled and not require user assembly for use.

8 COMMUNI-CATING

Chapter 8 Communicating

Communicating effectively with clients, collaborators, manufacturers and end-users is a vital core skill for any designer. It is crucial that you are able to communicate your ideas and thoughts to the other people involved in a design project by the use of design sketches, prototypes, presentation models or more formal oral presentations. Good communication is essential to successful product design projects, and engaging your clients, stakeholders and the general public in your work. This extends to the manner in which you present your work, how you present yourself, what you say and how you act. The following methods provide a number of ways to develop your presentation skills, from preparing a presentation to writing reports and maximizing the impact of presentation visuals and models.

PREPARING A PRESENTATION

The manner, style and procedure in which an individual designer researches and develops a design concept can often be quite distinctive and personal. It involves the internal thinking process and external drawing and prototyping processes, and so is necessarily a complex journey, often taking a non-linear route from A to B as a designer grapples with new research information, visual realizations of concepts and feedback from colleagues, clients and users. As a result, presenting the design development process in a clear and logical manner can be challenging. However, a designer must to be able to present his or her development work – and make explicit the concepts and reasoning behind certain design evaluations and decisions – in order to communicate effectively with others, both within and outside the studio. Common strategies to achieve this include keeping sketchbooks, research folders and journals, which can be edited on a rolling basis to provide evidence and justification for the decisions taken. These help form the foundation for preparing your presentation visuals.

Always try to arrive at a presentation five or ten minutes in advance, and aim to finish on time; don't attempt to say too much and then leave yourself with the need either to rush through or to go over the suggested limit. Over the page are ten top tips for a great presentation.

Fig. 1
Regular presentations within the design team ensure a smooth design development process.

TEN TIPS FOR A GREAT PRESENTATION

1. Structure your presentation
Move from the particular to the general before returning to the particular. Offer the viewer an immediate pay-off for listening – a 'visual hook' that will catch the audience's attention and demonstrate what the concept of the project is early on (usually on the first or second page). Generalize – don't go into every detail of everything you've done, just select the important parts and highlight those.

2. Show examples
Show visual examples throughout the presentation to demonstrate your concept.

3. Provide a take-away
It's useful to give people something to take away with them after a presentation, so they remember what the work was about and who you are. It is a good idea to supplement your presentation with handouts such as a postcard, a booklet or a small sample sheet.

4. Show models
If your proposal or aspects of your design and development work have involved 3D models and/or props, then have these with you so that they can be handed around while you are talking to the audience.

5. Never assume anything
Always expect that the audience does not know what is in your head and find a simple, effective way to talk them through the design project.

6. Avoid being too formal
You don't need to be overly formal to get your message across. Try to speak at a 'colleague' level of intelligence and in an approachable but professional style.

7. Sequence your presentation
Your presentation, and how you talk an audience through your project, should generally follow the following sequence:
— State the main problem/question/aim
— Show evidence of that problem and the need for the outcome
— Outline the development of a solution and the approach you've taken
— Provide the solution

8. Non-verbal communication
Non-verbal communication is as important as what you say and what's on the presentation sheets. Smile when talking, and look at who you're presenting to. This communicates enthusiasm and develops a rapport with the audience so that even if you forget the words or stumble, they will still be 'on your side'.

9. Be prepared
Practise, but don't try to memorize every detail and every word you want to say. It helps to have general notes for your presentation in the sequence you want to discuss things. It can be very stressful if you feel you need to recite a short essay. Practise with friends or colleagues so that they can give you feedback.

10. Content is king
Content is paramount. Always try to ensure the content is actually worth presenting and make sure the presentation is visually stimulating. It's easier to present if the audience is looking at the slides or pages and not at you all the time, so give them something good to look at.

Fig. 2
(Opposite) An informal presentation taking place at the Seren design studio in London. Seren employs a diverse range of creative employees from different disciplines such as product design, interaction design, sociology and computing science.

REPORT CREATION

By documenting your decision-making process, you will enable the creation of a readily understood archive of the rationale behind your design decisions. Such a report is useful for assimilating new team members and for quickly assessing the impact of changes as the product moves through design process towards manufacture and its launch in the marketplace.

Report writing is becoming increasingly commonplace in contemporary design projects and is an important skill for designers to acquire. Writing good design reports can be a real challenge, as they often need to reflect the final results of complex projects that have included detailed research, sometimes undertaken over a significant period of time. Usually, however, a lot of the detail must be left out of the main part of the report; instead the report should focus on the presentation of summary information, including the rationale behind critical design decisions leading to your overall recommendations. The detail can be placed in the appendices instead, and accessed later if needed.

A design report must comprise the following main levels of information:
— Executive summary
— Introduction
— Analysis
— Conclusion
— Appendices

The report should be well structured, stating your outcomes and recommendations clearly, using well-designed diagrams, charts and images, and providing the reader with a concise summary of the research conducted. A key element in a good design report is a clear structure. The following structural guidelines will help you create excellent reports every time.

1. Title page – this provides the title of the project, client name, date, your name and your organization.

2. Summary – this sets the problem in context, summarizes what you have done, and lists the key outcomes and recommendations.

3. Table of contents – this page clearly outlines each part of the report using section headings and page numbers.

4. Introduction – this both introduces and situates the problem being addressed, and discusses any previous research in the area.

5. Analysis section(s) (usually given specific titles) – this provides a walkthrough of the analyses that led to your overall recommendations. You should keep this simple throughout, using only essential diagrams, charts and images. Remember that the detail (e.g. raw data) should be placed in the appendices.

6. Conclusion – this should give a brief summary of what you have done, and include your recommendations.

7. References – this is a list in standard form of all the books, journals, magazines, scientific reports, websites and other resources you have referred to in the report.

8. Bibliography – this contains other books and resources you used during the study that might not be referenced explicitly in the report.

Chapter 8 Communicating

9. Appendices – you may have more than one appendix, which will describe in detail, if necessary, the analyses you have undertaken for the brief and the data you have obtained.

After you have written your design report, it is a good idea to go through the following questions to ensure you have covered everything:
— Have you answered the brief?
— Have you clearly understood the context of the problem?
— Have you fully articulated the analyses you have undertaken in the project?
— Have you clearly illustrated your results in diagrams, charts and images?
— Have you written the executive summary in clear and concise terms?
— Have you used appropriate and relevant vocabulary?
— Does the layout of your report clearly map the progression of your design project, including your research and results?

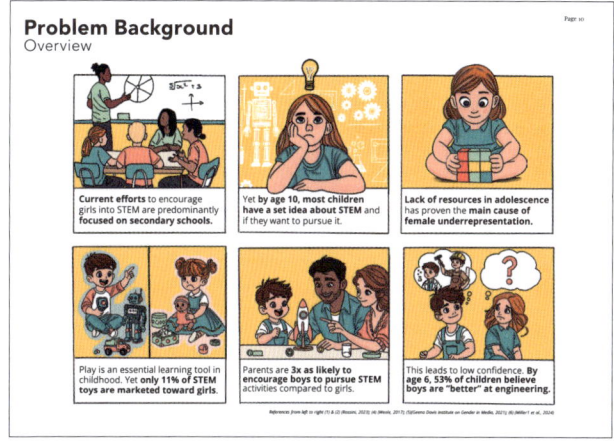

Fig. 3
Representative pages from a student design report by Eilidh Mathieson exploring creative ways to encourage more young girls to study STEM (Science, Technology, Engineering and Mathematics) subjects.

PRESENTATION VISUALS AND MODELS

During the development of a new product, a product designer will often make hundreds of quick sketches and models. However, when presenting these ideas to the client and others in order to communicate intentions of size, shape, scale and materials, these rough sketches and models will need to be tidied up so as to present something more visually seductive.

The designer has to shift from a three-dimensional idea into a two-dimensional sketch, and then back again into a three-dimensional representation of that idea. The widespread adoption of CAD has led to many designers developing their ideas virtually, visualizing their new products and concepts through photorealistic renderings of 3D models and scenes. The use of digital models allows for the quick exploration of new design ideas, materials and textures, and often provides a more cost-effective alternative when it comes to producing physical presentation models and photographing them.

However, while digital models are ideal for generating visuals for marketing and advertising material and presentations of new products, there is no substitute for a well-resolved tangible physical model that consumers and clients can see, feel and engage with. Product design is a three-dimensional discipline, and while the immediacy of marker renderings and the visual gloss and ease of CAD offer huge possibilities, it is essential that designers model their concepts physically and test and present them in the real world.

PRESENTATION GUIDELINES

When visualizing, modelling and presenting design concepts you should adhere to the following points:

— Visualize early – Don't just visualize as a presentation tool, but as a concept generation device that can convey your concepts clearly and concisely to as wide an audience as possible.
— Iterate often – Iterate as much as possible during the initial stages of the design process. This will help you generate ideas in a manner more conducive to evaluating a concept's merits rather than falling for the superficial qualities of a particular visual.
— Don't over-visualize – The aim of concept generation is to generate as many viable concepts as possible. Low-fidelity, rapid sketches and models are far more useful at this stage of the design process than more polished techniques, as they encourage debate.
— Visualize neutrally – When evaluating alternative design options it really helps to keep the quality and style of each visual or model as similar as possible. By presenting designs in a neutral manner, you can shed a sense of ownership, and the efforts of the entire team can be evaluated on a level playing field.
— Be aware of how people interpret visuals – You need to be fully aware of the subtle messages that different forms of visuals carry. For example, a rough pencil sketch has an immediacy that might imply an underdeveloped concept, while a photorealistic computer rendering may imply that what is, in fact, a mere concept is a finished design that is beyond criticism or change.

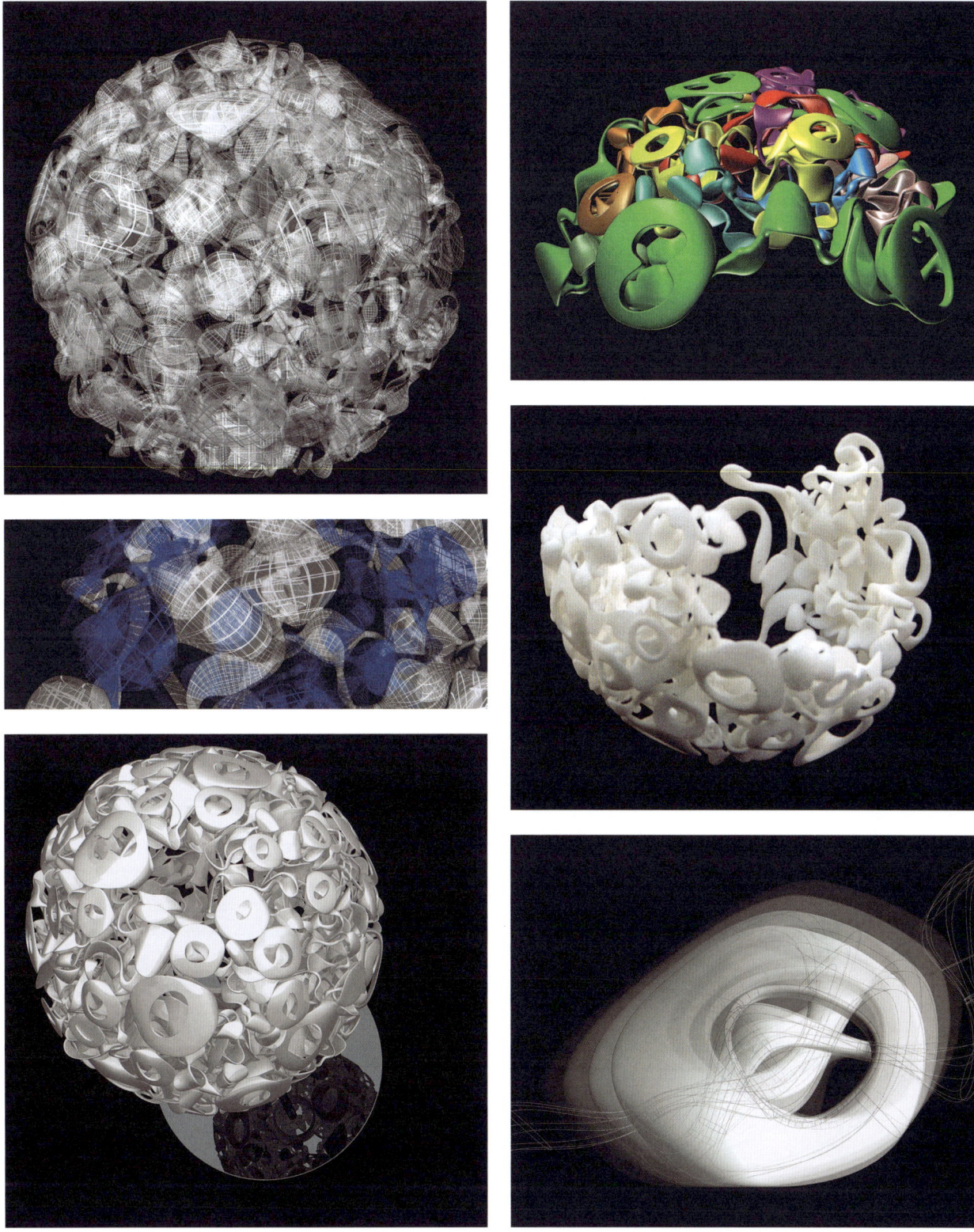

Fig. 4
Presentation visuals used to demonstrate the digital manufacture of the Entropia Light by Lionel T. Dean.

ENGAGING THE PUBLIC

Design research cannot reside in an academic ivory tower or insular design studio, and academic and professional design researchers are increasingly seeking to collaborate with and engage the public in their research activities, going on to share the benefits with the public.

Collaboration has become increasingly prevalent in design research, involving designers, researchers from various fields and stakeholders such as volunteers and community members. Collaboration and engagement are by definition a two-way process, involving dialogue and interaction with stakeholders, with the goal of generating mutually beneficial outcomes. Collaboration fosters dialogue and incorporates diverse perspectives, enhancing personal and professional growth while alleviating the isolation often associated with individual design research.

When initiating or joining a collaborative project, it is essential to clarify roles, hierarchies and working agreements to prevent implicit power dynamics from taking precedence. Discussions about power, privilege and individual capacities help acknowledge diverse skills and commitments among participants, allowing space for growth while recognizing differences in contributions.

Establishing clear agreements before commencing a design research project or activity enhances trust and co-creation, with initial discussions covering goals, media, target audiences and participants' expectations. Practical matters like budget, administration and time management must also be addressed, along with effective communication of responsibilities.

Public engagement can dramatically improve the quality and relevance of your research, helping you to refine your ideas and develop your presentation and communication skills. Those who engage with your research can play an invaluable role in contributing to that research, while stimulating their curiosity for your work and the products you are developing. Not only does the public raise relevant questions but projects that have been defined and researched in partnership with the public also often result in greater disciplinary and commercial impact and relevance.

Public engagement covers a range of different activities, from traditional one-way forms of engagement, such as public lectures and talks, to more interactive forms of engagement, such as participatory design. As a design researcher it is all too easy to lose perspective on why your research matters, especially when addressing longer-term speculative research that may only be commercialized in years to come. Discussing your work with the public can help you question your assumptions, introduce fresh perspectives to improve your thinking, and provide an opportunity to reflect on your design practice and research.

When determining how to present your research and engage the public, consider the following:

Purpose

Why do you want to engage the public with your research? This can be challenging to pin down, but it is often hard to determine a clear public engagement approach without knowing why you are doing it in the first place. Are you seeking to disseminate the results of your research? Are you hoping to encourage people to help you develop your design research further? Are you trying to promote your approach to design and the creative methods you employ? Are you aiming to consult the public on their views of your design research?

Audience
Think about who you are hoping to engage with. Who is your audience? How can you ensure you engage with them effectively? It is easy to think that your audience is an all-inclusive notion of the 'general public', but who are the key stakeholders in your design research? Once you have identified your audience, you should focus on understanding their interests and lifestyles. Why might they be interested in taking part in your public engagement activity? The more you understand your audience, the more successful your public engagement activity is likely to be.

Activity
What are you trying to do through engaging with the public? Be realistic about the resources you have at your disposal, and ensure that you are fully prepared to run the activity in an appropriate manner. There are lots of ways of engaging with the public, and you can present your work to audiences through a variety of forms, such as exhibitions, presentations, workshops, lectures, talks, blogs, forums and press activities.

Documentation
Thorough documentation of the collaborative and engagement process is vital, with all participants ideally contributing to the methods and goals of documentation and the publication strategies. Maintaining open dialogue throughout a research project or activity ensures that collaborators remain aligned regarding their aims and dissemination plans.

Assessment
Once you have undertaken a collaborative or engagement activity, evaluate what you have learnt from the process. Ensure that the aims of your public engagement are achievable, and make your objectives SMART: Specific, Measurable, Achievable, Relevant and Time-bound. Undertake ongoing evaluation (formative) to assess how successful your dissemination outputs such as exhibitions, workshops, blogs or papers were in engaging with your audience. This allows you to modify what you are doing. Finally, conduct summative evaluation at the end of the process to assess the success in achieving your outputs (results of your activity), outcomes (overall benefits) and the overall impact (effect and influence of the activity).

Communication skills are at the heart of public engagement, so you should adhere to the following guidelines:

- Adapt your presentation approach and content for different audiences.
- Listen to your audience and respond carefully to their questions and inputs.
- Respect and value your audience's contributions.
- Build on your audience's knowledge and understanding.
- Welcome feedback.
- Reflect on your own practice.
- Conduct formative and summative evaluation.
- Recognize when to seek advice or support from colleagues, key audience members and presentation experts.
- Be sensitive to issues of diversity and inclusion.
- Respect differences in understanding and attitudes.
- Be alert to social and ethical issues.
- Build and sustain effective partnerships.

CASE STUDY

AMO/REM KOOLHAAS ROADMAP 2050 PROJECT

Introduction
The work of Rem Koolhaas and OMA has won several international awards, including the Pritzker Architecture Prize in 2000, the RIBA Gold Medal (UK) in 2004 and the Mies van der Rohe European Union Prize for Contemporary Architecture in 2005. OMA's recent projects include the new headquarters for China Central and the Shenzhen Stock Exchange in China, De Rotterdam (the largest building in the Netherlands), the Zeche Zollverein historical museum in Germany, the Seoul National University Museum of Art and the much acclaimed Casa da Música in Portugal. The counterpart to OMA's architectural practice is AMO, a design and research think tank based in Rotterdam.

Objective
While OMA remains dedicated to the realization of buildings, AMO operates in areas beyond the traditional boundaries of architecture, including product design, media, politics, sociology, renewable energy, technology, fashion, curating, publishing and graphic design. AMO often works in parallel with OMA's clients to fertilize architecture with intelligence gleaned from this array of disciplines.

AMO has been involved in a number of high-profile renewable energy projects, including Zeekracht for the North Sea and the Energy Report for the World Wildlife Fund. In October 2009, European leaders committed to an 80 to 95 per cent reduction in CO_2 emissions by 2050. Roadmap 2050 was commissioned to determine how this goal could efficiently be met.

Methods
In design, the issue of sustainability is generally dealt with at the scale of products or buildings. Roadmap 2050 adopts a fundamentally different approach, seeking solutions that transcend the scale of a building, a city or even a nation. AMO developed a vision for an EU-wide de-carbonized power grid by 2050 through the creation of a report, proposing a practical guide to a prosperous, low-carbon Europe.

Results
The technical and economic analyses of the Roadmap 2050 report outline why a zero-carbon power sector is required to meet this commitment and illustrate its feasibility by 2050 given current technology. AMO contributed to the content development through the production of a graphic narrative about the geographical, political and cultural implications of a zero-carbon power sector. This narrative shows how through the complete integration and synchronization of the EU's energy infrastructure, Europe can take maximum advantage of its geographical diversity. If the Roadmap is followed, by 2050, the simultaneous presence of various renewable energy sources within the EU will create a complementary system of energy provision, ensuring energy security for future generations. Roadmap 2050 was commissioned by the European Climate Foundation, and the full report includes extensive technical, economic and policy analyses.
oma.com

Chapter 8 Communicating

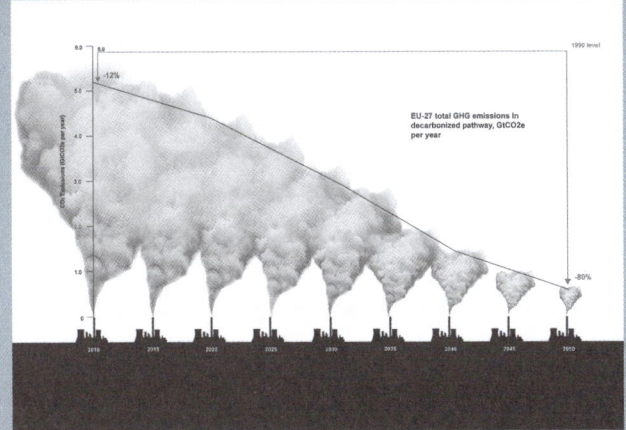
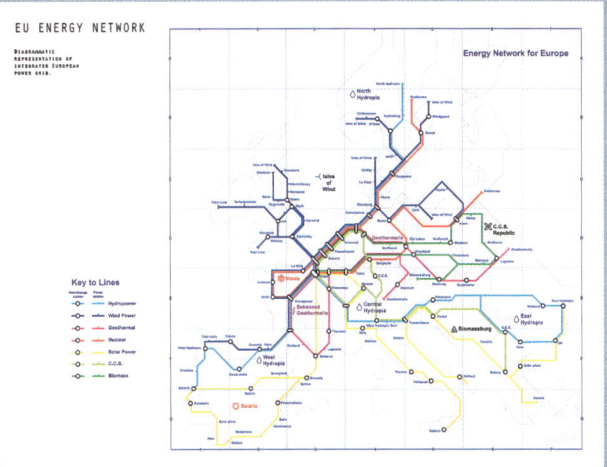

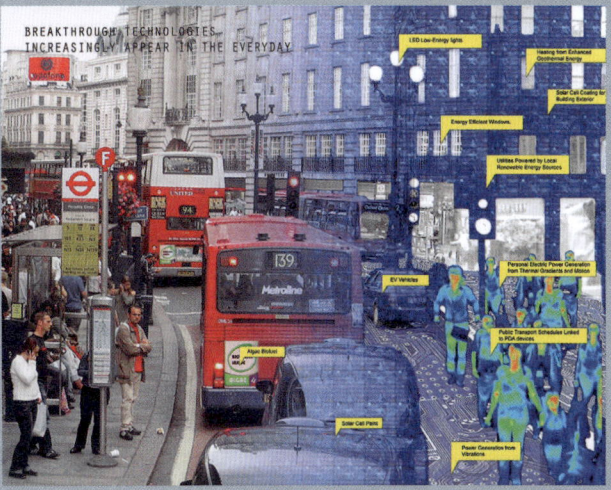
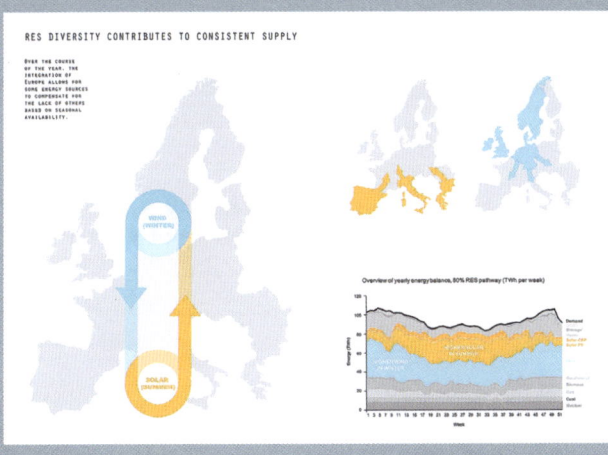

Conveying complex concepts is a key skill required by the contemporary designer. AMO adopted a diverse range of approaches to communicate their research findings for the Roadmap 2050 research project, from photo collage to information graphics, schematic diagrams and provocative renderings.

CASE STUDY

GIOVANNI INNELLA & GIONATA GATTO'S GEOMERCE

Introduction

GeoMerce is a speculative design project first presented in 2015 during Milan Design Week. It envisions a future where the scientific principle of phytomining – extracting metals from the soil using hyper-accumulator plants – is integrated into global financial systems. The project creates a thought-provoking scenario where agriculture evolves beyond food production to become a financial activity influenced by real-time metal-market trends. By blending elements of design, science and economics, GeoMerce challenges conventional perceptions of farming and highlights the complex entanglements of nature, technology and capital. Exhibited internationally, including at MoMA's *Broken Nature* exhibition, the project has sparked discussions on the implications of eco-capitalism and resource commodification.

Objective

GeoMerce explores and communicates the potential societal and environmental consequences of a future where farming is driven by financial speculation rather than traditional agricultural goals. Using speculative design as a form of future forecasting, the project examines the feasibility, risks and ethical considerations of such a scenario. By visualizing how hyper-accumulators could transform polluted land into financial assets, the project provokes reflection on the intersection of ecological restoration, economic exploitation and technological innovation. GeoMerce also seeks to engage diverse audiences by making the speculative accessible and inviting discussions on the possibilities and perils of green capitalism.

Methods

GeoMerce employs speculative design to forecast potential futures by creating a tangible and interactive installation. The project uses phytoextraction as a starting point, a scientific process where plants absorb metals from contaminated soil, accumulating them in their leaves. This concept is coupled with real-time data from the London Metal Exchange, where metal prices fluctuate based on global financial speculation. The integration of these elements creates a bio-digital system that visualizes the merging of ecological processes with economic mechanisms.

The installation is composed of five interconnected components:

1. Extraction units: hydroponic systems featuring hyper-accumulator plants, sensors and ion-selective electrodes track the extraction of metals in real time.

2. The brain: a bio-digital processor combines the extraction data with live metal prices, projecting the speculative financial value of the crops.

3. Circular plotters: three plotting units generate graphs that visualize the daily performance of metal extraction and market fluctuations.

4. The lenses: transparent leaves embedded in resin, showing accumulated metals, act as physical evidence of phytomining's microscopic processes.

5. GeoStory: a narrated animation projected onto a landscape-like surface illustrates a speculative timeline, from industrial pollution to the integration of farming and finance.

Below and opposite: Photos of the GeoMerce installation as exhibited during the Design Week in Milan, 2015.

Chapter 8 Communicating

Workshops, collaborations and dialogues with scientists, technologists and financial experts shaped the project's conceptual and technical framework. The design process also involved iterative prototyping and exhibition feedback to refine how speculative narratives were communicated.

Results

GeoMerce achieved its objective of provoking debate on eco-capitalism and the commodification of nature. The installation's interactive components enabled audiences to engage with complex ideas in an accessible and visually compelling manner. Visitors could observe how plants' extraction rates fluctuated in real time with market dynamics, blurring the boundaries between biology and economics.

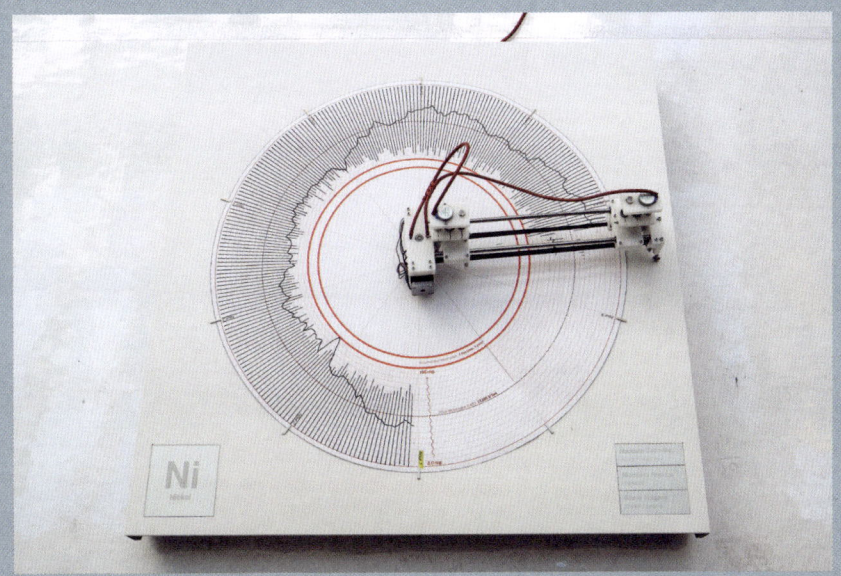

The project highlighted both the promises and pitfalls of a future where ecological processes become financial assets. On one hand, GeoMerce demonstrated how hyper-accumulators might be employed to rehabilitate polluted lands while generating economic value. On the other hand, it raised critical questions about the ethics of treating nature as a tradeable commodity. Could financial speculation incentivize overexploitation of phytomining? Would genetic modifications to improve extraction efficiency lead to further ecological imbalance?

GeoMerce exemplifies future forecasting by situating speculative scenarios within tangible, research-informed contexts. It underscores the power of design to envision possible futures and engage diverse stakeholders in meaningful dialogue about the trajectories of technology, ecology and society.

giovanniinnella.com
gionatagatto.com

TUTORIAL

How to create a great research presentation

Before you begin, think carefully about what you need to include in your presentation. If necessary, you might wish to clarify with your clients, collaborators and stakeholders exactly what is required of you and what facilities you can utilize. In terms of the visual aspects of your presentation, use websites, magazines, books, publications and other materials as a source for your layout designs. The following guidelines offer a number of suggestions on how to maximize the impact of your presentation and your presentation skills, whether you are creating digital presentations or physical presentations on paper:

Size
Whether you are presenting a keynote on a massive screen, creating for mobile devices or using hard copies, start by selecting the dimensions you will use for your presentation. Ensure any infographics such as charts and diagrams are legible.

Duration and volume
Consider how long you have to present, and what information you need to convey, and in what order. Scale the volume of your slides or pages to ensure you do not overrun a set time slot or require too much wall space.

Visual theme
A cohesive look with consistent typography and a simple colour scheme helps your audience stay focused on the content of your presentation.
    Create a few master pages to ensure your presentation looks professional and well designed. Add image and text frames to the master pages so you can drop in your content later without having to overthink the layout.

White space
Avoid cluttering your slides or pages with endless charts, tables or bullet points too small for anyone to read. Do not overload each slide or page with too many visuals or text; the most effective layouts are simple and allow some white space so crucial information stands out. Background colours are often difficult to handle and can easily make your work look less professional. White always works well.

Hierarchy
Make sure there is a variety of image and text sizes so that the important aspects stand out. Your key image should be noticeably larger than others so that it is the first thing a viewer looks at on the page. By making sure there is a hierarchy of information, you control how a viewer reads the slide or page.

Scale
When creating a hierarchy of information and imagery, consider using a set number of sizes for images and text throughout the entire presentation. This gives consistency and helps to create a look for all the pages. It also makes it much faster to put a presentation together. Create no more than three text styles so you can keep the title font, body font and footnote font consistent throughout the presentation. Set paragraph styles to change font and size with a click of a button. For example, all title text in Helvetica 20 point bold, all descriptive text in Helvetica 12 point regular and

Chapter 8 Communicating

Sample page from a student report by Nuria Mora Hurtado. A step-by-step approach ensures that the depth and breadth of research can be appreciated, evaluated and approved.

all caption text in Helvetica 9 point bold and italic. The same rules can be applied to the images, by making sure, for example, that all key images and smaller images are consistently sized.

Text
Choose one font for the entire presentation, but use different styles and sizes to indicate the hierarchy for titles, descriptive text and captions. A sans-serif font is one that has no flicks on the letters; these are often clean, modern and understated. Popular sans-serif fonts are Helvetica, Gill Sans and Futura.

Serif fonts have flicks on the letters; they are often classic, book-like and traditional. Popular serif fonts are Times New Roman, Baskerville and Garamond. There is a huge variety of fonts available, and this is very much a personal choice, but the ones listed above are a good starting point.

Never use white text on a black background, as it is difficult to read, particularly in a serif font at a small size. Never have a line length of more than 12 to 15 words – people will forget the last word of a line by the time their eye moves to the start of the next line, and so will not be able to absorb the information properly.

Format
Choose a format and template that suits your project and use it consistently for all the slides or pages in your presentation.

Images
Source material from the internet carefully. There is nothing worse than a small pixelated image in a presentation.

Editing
Be selective about what you show in a presentation; not everything needs to be used, and not all text is necessary for people to understand the concept or work. Allow time to put the presentation together, and then look at how much you are trying to include; have a colleague look at it with you and discuss whether you need less or more to communicate your ideas.

Incorporate video, audio or animations
Get your audience to pay more attention to you with animated presentations that feature video, audio and elegant transitions to break up static slides.

Add finishing touches
From films and sound clips to hyperlinks, cross-references and page transitions, consider using a range of interactive options to engage your audience.

Printing
When producing hard copies of a presentation always allow time to print a black and white version of the presentation pages to check how things look. Often what you see on the screen looks very different from what will be printed on a sheet of paper, particularly the scale of text and images. Also print out a test of any colour images to check that they are true to the tones you have chosen on-screen.

TUTORIAL

How to create a great research report

A research report is an account of the observations or study conducted by the report writer (researcher). A report can be written by almost anybody who can present his or her record of observations. However, there are several factors that will differentiate a good report from a bad one. Reports are only good if they reflect an accurate or faithful recollection of the events conducted, including the observation of people, products and processes, or information obtained from reading and/or past records. Bad and/or wrong data, such as manipulated or manufactured data, will result in a bad report.

Objectivity is another important characteristic in the quality of a report. A report must be an accurate account of the original information gathered; you must present the plain facts as you discovered them. Reporting, intentional or otherwise, that alters or modifies the nature of these facts would constitute bias or subjectivity and every effort must be made to produce an unbiased presentation.

A clear and well-organized format is a primary requirement of a good report. Often the specifications of the report will be stated by the recipient (e.g. the client, stakeholder, collaborator). However, as a general rule, nearly all forms of formal communication have the following main components.

Beginning (brief introduction)

First, introduce your study. State the origin and the rationale behind the idea (background). Support your rationale with previous studies, listing only the most significant and relevant sources (literature review). State clearly the main issues related to your study. These will form the focus of your report (objectives). State the benefits, relevance and advantages that will result from your study findings (significance of the study). Report how you conducted your study and describe your methodological strategy in tackling the issues (methodology) and the data collection procedures and instruments you used. For example, if you conducted a plain literature survey, include your plan and procedures for this with your gathered materials.

Middle (main section)

This section is the heart of your report and should contain all the main findings of your study (results). You should group similar findings under subheadings and support your statements with relevant previous studies or well-known facts and opinions. Remember, you are reporting the state of knowledge surrounding your study problem and you need to be accurate with every account. Be objective in your entire presentation of the results. Just present the facts that you have found, and leave the readers of your report some room for individual interpretation and judgement.

In subsequent sub-sections, present your own analyses, opinions and criticisms, supported by relevant literature and previous studies. You may like to title the heading of this section something like 'Results and discussions'.

End (concluding section)

Sum up the most important findings and insights in your report (summary) and reflect on whether you addressed all of the objectives in your study or not, stating briefly the limitations of your study. Finish your report with recommendations that leave readers with precise answers to your questions. Also, list any implications and applications of the study, and suggest where a subsequent researcher might go from here. This should include a section where you cite formally all your sources of information (references) so that readers of the report have the opportunity to verify or redo your study. Finally, you may have to include an section where you can attach any additional information or relevant materials – such as tables, graphs, photos or CDs – that you have used in the report (appendices).

Initial Prototyping

Objective
To create basic initial models to better visualise and investigate the geometry and function of the concepts. Specifically, how the panel pulls out and the concertina mechanism connects the panels.

Process
Cardboard and card were the materials predominantly used to model this concept, initially cutting out the two panels and adding a fan to the front panel with a small shaft so that it could turn. The panels were then connected with four strips of card folded into concertinas to aid the pulling out of the product.

A metal bar and two hollow plastic tubes were then nested together to mimic a telescopic pole which was fixed in the centre of the two panels as the hanging rail. Cardboard and lengths of string were then used to model the position and basic function of the pull-out hanging bars. Finally, the tissue paper was folded and added to represent the top section of the cover.

Key Insights
The concertina mechanism used in the four corners for the frame may not be the best way to facilitate the movement of pulling out the front panel. It would require a number of hinges which would take time to assemble and may not be strong enough themselves to support the movement. If the concertina mechanism was kept it would have to be developed to include runners for each segment similar to those found on Bi-fold doors.

The hanging bars which are pulled out from the back panel will also require further development. The mechanism must enable the bars to be pulled out fully while supporting the weight of the laundry as well as fold away entirely. This mechanism will require further prototyping individually which is more detailed.

Above and below: These representative pages by Lauren Keith demonstrate the range of approaches and depth required for a typical research report. A coherent graphical style ensures that the reader can focus on the content.

Developing Aesthetics
The Cover

Objective
To experiment with concertina type folds that could be used to increase the visual intricacy of the protective cover.

Process
The cover used within the product has one primary function, to protect the laundry that is drying from unpredictable adverse weather conditions. Since the product extends and contracts to save space when not in use, the cover must also have the capability of compressing.

In the initial sketches and communications of the design the cover took the form of a basic concertina. It opened out to full length and appeared flat and then could also fold back tightly. The cover will be pulled across the entire product by the user when the product is in use making it a very prominent feature. For this reason it was decided that the folds of the cover could be manipulated and made more interesting.

Inspiration for the how the form could be conveyed came from **"Folding Techniques for Designers"** by Paul Jackson. In this book the writer illustrates a number of folding techniques and patterns to create a vast range of shapes from sheet material into 3D forms.

From the range of techniques five were selected from the Spans and Parabolas section (Jackson, 2011, p.170) and prototyped as they were of a similar form to the required shape of the cover and appeared as though they could contract. The forms were created by marking, scoring and then folding card in the Patterns provided in the book. Handling the forms physically mean that they could be analysed, understood and evaluated at a deeper level.

The five cover prototypes can be seen illustrated so they can be compared in size and shape. Each are individually discussed on the subsequent page.

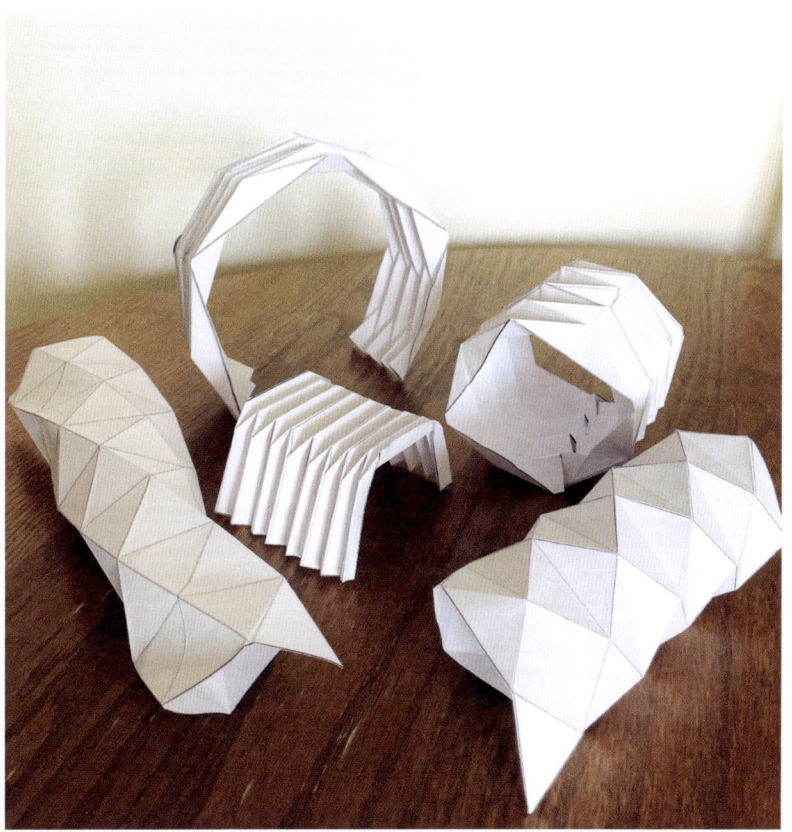

SUMMARY

This book has introduced a rich and diverse range of research methods and tools, along with ideas of how and when to deploy these methods effectively – all of which can be used to inform your design thinking and doing.

Each method will contribute a firm foundation for your design research, while the tutorials and real-world examples provided in the case studies will encourage you to develop your own research approach and toolkit. Here is a summary of the key points of the book, and of product design.

Looking

Design research usually starts with looking – observing the world around you, and using research methods to discover what people really need, want and do rather than just what they say they need, want and do. Observing and examining people's emotional attachments to their belongings, and carrying out the forensic analysis of designed products with a critical eye, can also help you to design and develop better products.

Learning

Designers can then learn from what people really need, want and do by using effective techniques such as precedent analysis, role playing and 'try it yourself'. Adopting these methods will enable you to learn first-hand what using a particular product in a specific context feels like.

Asking

A variety of straightforward ways of obtaining information from users, such as questionnaires, focus groups, interviews and the creation of personas, will rapidly provide you with an understanding of the multifaceted relationships that exist between users and the designed products and services they use every day.

Making

You can then move into the creation of models and prototypes to help inform your design and decision-making processes, communicate your design concepts, and enable users and clients to understand and explore how they might like to engage with the planned product or service.

Testing

A range of testing methods, including user trials, test rigs and safety testing, will then aid informed decision-making, and ensure a streamlined design and development process that avoids costly mistakes or delays.

Selecting & Evaluating

Selecting the wrong product design proposal to develop can be very costly to manufacturers and stakeholders in terms of time, money and other valuable resources – the numerous systematic methods for evaluating and selecting design proposals described will help maximize your chances of selecting the right one.

Communicating

The final stage of the design research process is communication. It is essential for a designer to be able to communicate clearly with potential customers, clients and stakeholders. Ideas and thoughts can be disseminated to the various people involved in a design project through the use of design sketches, prototypes, presentation models and other more formal presentations. Good communication is key to successfully presenting product design research and practice. By using a number of the established and emergent research methods contained within this book, you can help ensure you develop your designs in a carefully considered, transparent, iterative process.

Conclusions

Design research employs both qualitative and quantitative research methods, including ethnography, mapping, trend forecasting, cultural comparisons, video diaries, probes and a host of other methods to collaboratively shape our local and global futures. Design projects routinely transcend economic, national and cultural borders, and this increasingly challenging environment requires that product designers understand different cultures, evaluate design proposals sensitively and communicate with people from very different backgrounds.

New technologies are transforming how we communicate, learn and make, but these can also make a designer's life, and indeed the lives of everybody, more complicated.

We are in a decisive decade that demands more inspired, informed and engaged designers who can design products, services and systems that create preferable futures and help rebalance the present.

The implications for product designers today, therefore, run to some very serious social, cultural and environmental issues. It is hoped that this book will inspire you to research and develop future products that are sensitive to these issues, and to help shape a future world worth living in.

GLOSSARY

appearance model
A model created to simulate the look and physical characteristics, rather than the function, of a proposed product.

bodystorming
A research method in which a design team attempts to physically recreate a situation through role play, in order to better imagine its social and physical considerations.

boundary user
A user on the limit of being able to use a product, a useful consideration in identifying opportunities for design improvement.

brand DNA analysis
Multilayered analysis of a branded product, integrating quantitative data and qualitative research from many sources.

brandscaping
A structured evaluation of an industry sector achieved by selecting the products most representative of each brand, and comparing their most characteristic elements and important features.

CAD
Computer-aided design, also computer-aided drafting. The use of computer systems to assist in the creation, modification, analysis or optimization of a design.

circular design
The practice of creating durable, reusable, repairable and recyclable products that generate zero waste to support a circular economy.

clay model
A product model literally made of clay or industrial plasticine; also a generic term for **appearance models** of any material, as these were frequently made of clay in the past.

CMF design
Colour, Materials and Finish is an area of product design that focuses on the chromatic, tactile and decorative identity of products.

co-design
A collaborative approach where designers work together with non-designers to create solutions.

competitor product analysis
Examining and evaluating a product and its competitors with respect to a predefined set of both qualitative and quantitative criteria.

concept map
A graphic illustration in which two or more concepts are linked by words that describe their relationship. Developed by Professor Joseph D. Novak at Cornell University in the 1960s.

critical design
Speculative design proposals challenge assumptions and conceptions about the role products play in everyday life.

crowdsourcing
Outsourcing tasks traditionally performed by experts to a group of people through an online open call.

crowdvoting
A form of **crowdsourcing** that makes use of the power of popular opinion and doubles as a marketing tool: companies ask users to vote ideas up or down in real time, simultaneously generating useful feedback and consumer interest.

cultural comparisons
The careful use of personal or published accounts to reveal differences in national or cultural behavioural traits, which will have design implications when producing for global markets.

cultural probe
A means of gathering information in a creative manner with minimal intrusion: subjects are provided with items such as cameras, voice recorders and notebooks, and asked to record and report back on their experiences over a period of time.

data analytics
Converts raw data into actionable insights. It includes a range of tools, technologies and processes used to find trends and solve problems by using data.

data visualization
The representation of data through the use of charts, infographics and animations. These visual displays of information communicate complex data relationships and data-driven insights in a way that is easy to understand.

a day in the life
An intensive research method that aims to provide a representative snapshot of the life of a potential product user, and may reveal issues inherent in their routines and circumstances.

digital ethnography
The use of digital tools such as cameras, computers and the internet to accelerate the process of data collection, analysis and presentation in **ethnography**.

discursive design
The creation of products whose primary purpose is to communicate ideas and encourage discourse and social debate.

drop testing
Measuring the durability of a part or material by subjecting it to a free fall from a predetermined height on to a surface, under prescribed conditions.

empathy tools
Physical or software devices that designers can use to reduce their ability to interact with a product, and thereby gain an impression of the experiences of users with disabilities or special conditions. Also known as 'capability simulators'.

end-user
The person who ultimately uses a product, as opposed to others who may handle it or purchase it within a supply chain.

ethnography
The study of the culture, knowledge, language, values and system of meanings of a people or group.

experience prototyping
Any kind of non-physical representation of a product, which itself might be a service, lifestyle or experience, that enables a design team to learn from a simulation of the proposed product's use in different contexts.

extreme user
An individual who is either extremely familiar or completely unfamiliar with a particular product, service or system.

failure modes and effects analysis (FMEA)
A structured risk analysis tool to identify, quantify and mitigate the specific risks associated with individual assemblies or components by determining the location and nature of a failure.

focus group
A number of people brought together in one place to discuss a particular issue or set of issues, such as their experiences of a particular product, service or system.

future forecasting
The identification of future aesthetic preferences as well as the prediction of global ecological, technological and financial changes, to help designers anticipate future needs.

generative AI
Generative AI models produce text, images, videos or other forms of data. These models learn the underlying patterns and structures of their training data and use them to produce new data based on inputs such as prompts.

image board *see* **mood board**

inclusive design
The design of mainstream products and services that are accessible to, and usable by, as many people as reasonably possible without the need for special adaptation or specialized design.

journey mapping
A visualization document that showcases the steps that a user takes in a process to accomplish a particular goal, such as purchasing and using a product.

literature review
A text considering the critical points of current knowledge on a topic, including substantive findings as well as the theoretical and methodological. A detailed form of **secondary research**.

market positioning
The decisions made by an organization about how they want users to see their product relative to the existing market or **brandscape**.

matrix evaluation
A quantitative technique used to evaluate concept design proposals by ranking them against set criteria stated in the **product design specification (PDS)** or other concept design proposals. Sometimes referred to as 'Pugh's Method' after its creator, Professor Stuart Pugh.

metrics
The set of qualitative features chosen by researchers to be measured in product usability tests.

mind map
A visual representation of hierarchical information, made popular by the psychologist Tony Buzan, and intended to represent words and ideas in a way that engages both sides of the brain.

mock-up
An easily fabricated, life-size physical model constructed from rough materials, used to evaluate the physical interaction, scale and proportion of product design concepts.

mood board
A collaged board containing images, text and samples of objects, used by designers to portray a range of potential directions for a specific product or brand, to help develop and communicate design concepts. Also known as an 'image board'.

mystery shopper
A researcher posing as a regular consumer, whose job it is to carry out market research or internal quality control with their identity concealed.

name swapping
A market research technique, which involves swapping the names and logos on different product designs from the same market, and discussing if and why the resulting designs are 'wrong' for the branded products being researched.

pen portrait *see* **persona**

perceptual mapping
Arranging market research data on an X–Y axis to visually compare attributes such as the perceived cost, quality and impact of a brand or product. Used to evaluate corporate design positioning strategy.

persona
A fictional character, based on real-life observations of archetypal users, created to represent groups of users within a targeted demographic who might all use particular products, brands and services in a similar way. Also known as a 'pen portrait'.

primary research
Original research undertaken by the design researcher. More time-consuming and costly than **secondary research**.

product autopsy
The analysis of a product that has reached the end of its life, often involving a full disassembly, to discover how each component has fared, as well as to assess the design in more general terms.

product camouflage
A research method involving the modification of a series of existing designs, each with different elements removed. Through **focus group** discussions, the value of each element can be determined.

product champion
A member of the new-product design and development team who supports the instigation and development of concept design and acts as a go-between linking various departments within an organization, becoming an expert on that particular product.

product collage
A **mood board** created by potential users as part of design research, to help them articulate complex ideas and perceptions related to a design issue.

product design specification (PDS)
A document that clearly sets out the parameters that must be met by a design, presented initially before a project is taken on, but often adapted throughout the design process.

proof of concept model
Mock-ups that do not incorporate any product styling and are only intended to demonstrate the basic mechanism of a product, as proof of its potential viability.

Pugh's Method *see* **matrix evaluation**

qualitative research
Research into consumer or user behaviour that focuses on questions of *why* and *how* decisions or actions are made, often using smaller but more focused sample groups.

quantitative research
Research into consumer or user behaviour focusing on objective questions of *what*, *where* and *when* decisions or actions are made, ideally using large sample groups and producing results that can be displayed as graphs or statistics.

quick-and-dirty prototype
Quickly built prototypes using rough materials for speed, which are used to communicate a concept to other members of a design team. Also known as 'rough-and-ready prototype'.

rapid ethnography
An approximate form of **ethnography**, used when quick results are required.

rapid iterative testing and evaluation (RITE)
A form of product usability testing that encourages testers to 'think aloud' (**user narration**), enabling the designer to step in and change the user interface of a product the moment a problem has been identified and a rapid solution has been devised.

rapid prototyping
The automatic construction of detailed physical objects from computer data using a range of 3D printing technologies.

regenerative design
An approach in which products are designed to co-exist and co-evolve with nature over time.

role playing
A process in which designers assume the role of the various stakeholders in a project and act out scenarios to gain a better understanding of important issues.

rough-and-ready prototype *see* **quick-and-dirty prototype**

sampling
Selecting a subset of individuals, objects or items for study, intended to be representative of the wider group, from a population or marketplace.

scenario testing
The creation of future scenarios, demonstrating speculative products being used by ordinary people in realistic future contexts, presented through storyboards, texts, photography, film and plays, which can help product designers communicate and evaluate design proposals within their intended context.

schematic sketches
Sketches describing the fixed dimensional parameters of a design, including vital data such as off-the-shelf components to be used and ergonomic considerations, with less emphasis on appearance.

secondary research
The summary, collation and/or synthesis of existing research.

semiotics
The study of the use and understanding of signs within a particular culture.

shadowing
A method in which a researcher closely follows an individual or small team over an extended period of time throughout their working day.

sketch model
Generally crude full-size or scale models that aim to capture the embryonic ideas emerging from the design team's initial concept development; often an early attempt to realize a design in three dimensions before proceeding with more detailed work.

speculative design
An approach to design that focuses on imagining future scenarios and possibilities.

stereolithography
An additive manufacturing process used in 3D printing, which builds up layers of solid plastic using an ultraviolet laser and a liquid resin vat. Used for **rapid prototyping**.

task analysis
A systematic method for understanding the steps, decisions and actions involved in completing a specific task.

test rig
A **mock-up** that replicates a mechanical action or enables a physical property of a design, such as its strength, stiffness, comfort or durability, to be tested.

thematic sketches
Intentionally fluid, dynamic and expressive exploratory sketches, which convey a product's physical form, characteristics and overall aesthetic.

think aloud protocols see **user narration**

touchpoint
The means by which a user comes into contact with a product throughout its lifecycle, including interactive internet features, printed documents, physical devices, retail outlets and call centres.

touchpoint wheel
A tool for analysing the entire customer journey, which summarizes all the points of interaction where a customer can be intentionally or unintentionally influenced.

trend forecaster
A specialist working within a market research or similar organization who postulates possible future trends in fashion, culture and technology by evaluating past and present trends.

trend spotting
Detailed analysis of new design, commercial, visual or fashion trends as they begin to appear, looking at their qualities and probable development.

try it yourself
The act of using a prototype, existing or new product as part of the research process in order to gain first-hand experience of how it performs and how it feels to use it.

type testing
Benchmark tests, particular to a specific type of product, normally laid out by an independent quality-control organization in order to provide fair and balanced comparison.

unfocus group
A method for gaining a number of interpretations on a given design problem employing diverse individuals in a workshop-style setting contributing to concept design generation or evaluation activities.

user narration
A method for identifying users' concerns when using specific products, systems and services by asking them to think about and describe aloud their experience as they engage with or use them. Sometimes referred to as 'think aloud protocols'.

user trials
Trials of either new or existing products carried out by groups of users under controlled conditions. These often use prototypes as a cheaper alternative to field trials with a finished product.

RESOURCES

Further reading

Albrecht, D. et al., *Design Culture Now*, Laurence King Publishing, London, 2000

Antonelli, P., *Supernatural: The Work of Ross Lovegrove*, Phaidon Press, London, 2004

Antonelli, P., *Humble Masterpieces: 100 Everyday Marvels of Design*, Thames & Hudson, London, 2006

Antonelli, P. and Aldersey-Williams, H., *Design and the Elastic Mind*, Museum of Modern Art, New York, 2008

Bramston, D. and YeLi, *Idea Searching for Design: How to Research and Develop Design Concepts*, Bloomsbury Visual Arts, London, 2019

Brown, T., *Change by Design: How Design Thinking Creates New Alternatives for Business and Society*, Harper Business, New York, 2019

Buchanan R., *Design Research and the New Learning*, Bloomsbury, New York, 2018

Cagan, M., *Inspired: How to Create Products Customers Love*, O'Reilly Media, Sebastopol, California, 2017

Chia, P., *Design Incubator: A Prototype for New Design Practice*, Laurence King Publishing, London, 2018

Cooper, R. and Mullagh, L. (eds), *Design and Covid-19: From Reaction to Resilience*, Bloomsbury Visual Arts, London, 2024

Cooper, R. and Press, M., *Design Management: Managing Design*, John Wiley & Sons, London, 1995

Cross, N., *Designerly Ways of Knowing*, Springer, London, 2006

Cross, N., *Engineering Design Methods: Strategies for Product Design* (5th edn), John Wiley & Sons, Hoboken, New Jersey, 2021

Crouch, C. and Pearce, J., *Doing Research in Design*, Berg, London, 2012

Curedale, R., *Design Research Methods: 150 Ways to Inform Design*, Design Community College Inc., Los Angeles, 2013

Dixon, T. et al., *And Fork: 100 Designers, 10 Curators, 10 Good Designs*, Phaidon Press, London, 2007

Dormer, P., *The Meanings of Modern Design*, Thames & Hudson, London, 1991

Dormer, P., *Design since 1945*, Thames & Hudson, London, 1993

Dreyfuss, H., *Designing for People*, Simon and Schuster, New York, 1955

Dunn, N. et al. (eds), *Flourish by Design*, Routledge, Abingdon, 2024

Dunne, A., *Hertzian Tales: Electronic Products, Aesthetic Experience, and Critical Design*, MIT Press, Cambridge, Mass., 2008

Dunne, A. and Raby, F., *Speculative Everything: Design, Fiction, and Social Dreaming*, MIT Press, Cambridge, Mass., 2013

Dunne, A. and Raby, F., *Design Noir: The Secret Life of Electronic Objects* (2nd edn), Bloomsbury Visual Arts, London, 2021

Escobar, A., *Designs for the Pluriverse: Radical Interdependence, Autonomy, and the Making of Worlds*, Duke University Press, Durham, North Carolina, 2018

Fairs, M., *Twenty-first Century Design*, Carlton Books, London, 2006

Fiell, C. and Fiell, P., *Design Handbook (Icons)*, Taschen, Cologne, 2006

Fiell, C. and Fiell, P., *Design Now: Designs for Life – From Eco-design to Design-art*, Taschen, Cologne, 2007

Fiell, C. and Fiell, P., *Design of the 20th Century* (new edn), Taschen, Cologne, 2022

Flood, C. and Grindon, G. (eds), *Disobedient Objects*, V&A Publishing, London, 2014

Fuad-Luke, A., *Design Activism: Beautiful Strangeness for a Sustainable World*, Earthscan, London, 2009

Fukasawa, N., *Naoto Fukasawa*, Phaidon Press, London, 2007

Fukasawa, N. and Morrison, J., *Super Normal: Sensations of the Ordinary* (2nd edn), Lars Muller Publishers, Baden, Switzerland, 2007

Fulton Suri, J., *Thoughtless Acts?*, Chronicle Books, San Francisco, 2005

Gaver, W. et al., 'Design: Cultural Probes and the Creative Process', in *The Routledge Handbook of Design Research Methods*, Routledge, New York, 2018

Gay, P. et al., *Doing Cultural Studies: The Story of the Sony Walkman* (2nd edn), SAGE Publications, London, 2013

Goodman, N., *Ways of Worldmaking*, Hackett Publishing Company, Cambridge, Mass., 1978

Goodwin, K. and Cooper, A., *Designing for the Digital Age: How to Create Human-Centered Products and Services*, Wiley, Indianapolis, 2009

Groß, B. and Mandir, E., *Designing Futures: Speculation, Critique, Innovation*, Laurence King Publishing, London, 2024

Gunn, W. et al., *Design Anthropology: Theory and Practice*, Bloomsbury Academic, New York, 2013

Hallgrimsson, B., *Prototyping and Modelmaking for Product Design*, Laurence King Publishing, London, 2012

Hammersley, M. and Atkinson, P., *Ethnography: Principles in Practice* (3rd edn), Routledge, Abingdon, 2007

Hara, K., *Designing Design*, MIT Press, Cambridge, Mass., 2015

Henry, K., *Drawing for Product Designers*, Laurence King Publishing, London, 2012

Heskett, J., *Toothpicks and Logos: Design in Everyday Life*, Oxford University Press, Oxford, 2002

Hudson, J., *1000 New Designs and Where to Find Them: A 21st Century Sourcebook*, Laurence King Publishing, London, 2006

Hudson, J., *Process: 50 Product Designs from Concept to Manufacture*, Laurence King Publishing, London, 2008

Ingold, T., *Making: Anthropology, Archaeology, Art and Architecture*, Routledge, London, 2013

Johnson, S., *Emergence: The Connected Lives of Ants, Brains, Cities, and Software*, Penguin Books, London, 2001

Jordan, P., *Designing Pleasurable Products*, Taylor and Francis, London, 2002

Julier, G., *The Culture of Design* (3rd edn), SAGE Publications, London, 2014

Kalantidou, E. and Fry, T. (eds), *Design in the Borderlands*, Routledge, Abingdon, 2014

Kelley, T., *The Ten Faces of Innovation*, Profile Books, London, 2006

Koskinen, I., et al., *Design Research Through Practice: From the Lab, Field, and Showroom*, Morgan Kaufmann, Burlington, Mass., 2011

Lefteri, C., *Materials for Inspirational Design*, Rotovision Publishers, Hove, 2006

Lefteri, C., *Making It: Manufacturing Techniques for Product Design*, Laurence King Publishing, London, 2007

Lefteri, C., *Materials for Design*, Laurence King Publishing, London, 2014

Malpass, M., *Critical Design in Context: History, Theory, and Practice*, Bloomsbury, London, 2017

Mattelmäki, T., *Design Probes*, University of Art and Design, Helsinki, 2006

McKercher, K.A., *Beyond Sticky Notes: Doing Co-design for Real: Mindsets, Methods and Movements*, 2020

Moggridge, B., *Designing Interactions*, MIT Press, Cambridge, Mass., 2006

Morris, R., *The Fundamentals of Product Design*, AVA Publishing, Lausanne, 2009

Morrison, J. and Schelbert, C., *Everything but the Walls*, Lars Muller Publishers, Baden, Switzerland, 2006

Morton, T., *Hyperobjects: Philosophy and Ecology after the End of the World*, University of Minnesota Press, Minneapolis, 2013

Muratovski, G., *Research for Designers: A Guide to Methods and Practice*, SAGE Publications, London, 2016

Osterwalder, A. et al., *Value Proposition Design: How to Create Products and Services Customers Want*, Wiley, Hoboken, New Jersey, 2014

Oswald, L.R., *Doing Semiotics: A Guide for Marketers at the Edge of Culture*, Oxford University Press, Oxford, 2020

Papanek, V., *Design for the Real World: Human Ecology and Social Change*, Pantheon Books, New York, 1971

Parsons, T., *Thinking: Objects – Contemporary Approaches to Product Design*, AVA Publishing, Lausanne, 2009

Pink, S., *Doing Visual Ethnography: Images, Media and Representation in Research* (3rd edn), SAGE Publications, Los Angeles, 2013

Polanyi, M., *The Tacit Dimension*, University of Chicago Press, Chicago, Illinois, 1966

Potter, N., *What Is a Designer: Things, Places, Messages* (4th edn), Hyphen Press, London, 2008

Proctor, R., *1000 New Eco Designs and Where to Find Them*, Laurence King Publishing, London, 2009

Pye, D., *The Nature and Aesthetics of Design*, A&C Black Publishers, London, 2000

Rasch, M. et al., *Hands on Research for Artists, Designers and Educators*, Set Margins, 2024

Rawsthorn, A. and Antonelli, P., *Design Emergency: Building a Better Future*, Phaidon, London, 2024

Rodgers, P., *Inspiring Designers*, Black Dog Publishers, London, 2004

Rodgers, P., *Little Book of Big Ideas: Design*, A&C Black Publishers, London, 2009

Rodgers, P., *Design for People Living with Dementia*, Routledge, New York and London, 2022

Rodgers, P. and Milton, A., *Product Design*, Laurence King Publishing, London, 2011

Rodgers, P. and Yee, J. (eds) *The Routledge Companion to Design Research* (2nd edn), Routledge, Abingdon and New York, 2023

Rowe, P.G., *Design Thinking*, MIT Press, Cambridge, Mass., 1987

Sanders, E.B.-N. and Stappers, P.J., *Convivial Toolbox: Generative Research for the Front End of Design*, BIS Publishers, Amsterdam, 2013

Schön, D.A., *The Reflective Practitioner: How Professionals Think in Action*, Basic Books, New York, 1984

Schön, D.A., *Educating the Reflective Practitioner*, Jossey-Bass, San Francisco, California, 1987

Slack, L., *What Is Product Design?*, Rotovision Publishers, Hove, 2006

Stickdorn, M. and Schneider, J., *This Is Service Design Thinking: Basics, Tools, Cases*, Wiley, Hoboken, New Jersey, 2014

Sudjic, D., *The Language of Things*, Allen Lane, London, 2008

Swan, S., *Design Research: Methods and Perspectives*, Fairchild Books, London, 2019

Thackara, J., *Design after Modernism: Beyond the Object*, Thames & Hudson, London, 1988

Tharp, B. and Tharp, S., *Discursive Design: Critical, Speculative, and Alternative Things*, MIT Press, Cambridge, Mass., 2022

Thompson, R., *Manufacturing Processes for Design Professionals*, Thames & Hudson, London, 2007

Troika et al., *Digital by Design: Crafting Technology for Products and Environments*, Thames & Hudson, London, 2008

Ulrich, K.T. et al., *Product Design and Development* (7th edn), McGraw Hill Education, New York, 2020

Vaughan, L., *Practice-based Design Research*, Bloomsbury, London, 2017

Verganti, R., *Design-driven Innovation*, Harvard Business School, Boston, Mass., 2009

Yablonsky, S., *Product Design Essentials: 100 Principles for Creating and Refining Designs*, Rockport Publishers, Gloucester, Mass., 2015

Design competitions

Braun Prize
braunprize.com

Design & Art Direction (D&AD), Student Awards
dandad.org

Good Design Award
g-mark.org

IF Concept Award
ifdesign.de

Materialica Design Award
materialicadesign.com

Muji Award
muji.net/award

Promosedia International
promosedia.it

Red Dot Design Award
red-dot.de

Royal Society of Arts
Design Directions Student Awards
rsa-design.net

Useful addresses

Australia

Powerhouse Museum
500 Harris Street, Ultimo
PO Box K346, Haymarket
Sydney, NSW 1238
powerhousemuseum.com

Belgium

Design Museum
Jan Breydelstraat 5
9000 Ghent
design.museum.gent.be

Canada

DX: The Design Exchange
234 Bay Street, PO Box 18
Toronto Dominion Centre
Toronto, ON M5K 1B2
dx.org

Denmark

The Danish Museum of Art & Design
Bredgade 68/1260, København K
kunstindustrimuseet.dk

Germany

Bauhaus-Archiv Museum of Design
Klingelhöferstrasse 14
D-10785 Berlin
bauhaus.de/english/index.htm

Die Neue Sammlung
Barer Strasse 40
80333 Munich
die-neue-sammlung.de

Red Dot Design Museum
Gelsenkirchener Strasse 181
45309 Essen
en.red-dot.org/371.html

Vitra Design Museum
Charles-Eames-Str. 1
D-79576 Weil am Rhein
design-museum.de

Ireland

Institute of Designers Ireland
WeWork, 2 Dublin Landings,
North Dock, Dublin, D01 V4A3
idi-design.ie/

Mexico

Mexican Museum of Design
Av. Francisco I. Madero No. 74
Colonia Centro Histórico
Delegación Cuauhtémoc
México D.F, C.P. 06000
mumedi.org

Singapore

Red Dot Design Museum
28 Maxwell Road
Singapore, 069120
www.red-dot.sg/concept/museum/main_page.htm

United Kingdom

Chartered Society of Designers
1 Cedar Court, Royal Oak Yard
Bermondsey Street
London, SE1 3GA
csd.org.uk

Design Museum
224–238 Kensington High Street
London, W8 6AG
designmuseum.org

Museum of the Home
Kingsland Road
London, E2 8EA
museumofthehome.org.uk

Institution of Engineering Designers
Courtleigh, Westbury Leigh, Westbury
Wiltshire, BA13 3TA
ied.org.uk

Victoria and Albert Museum
Cromwell Road
London, SW7 2RL
vam.ac.uk

United States

Cooper-Hewitt, National Design Museum
2 East 91st Street, New York
NY 10128
cooperhewitt.org

The Eames Office
850 Pico Boulevard, Santa Monica
CA 90405
eamesoffice.com

Industrial Designers Society of America
45195 Business Court, Suite 250,
Dulles
VA 20166-6717
idsa.org

Museum of Arts and Design
2 Columbus Circle, New York
NY 10019
madmuseum.org

Museum of California Design
PO Box 361370, Los Angeles
CA 90036
mocad.org

Museum of Design, Atlanta
285 Peachtree Center Avenue
Marquis Two Tower, Atlanta
Georgia 30303-1229
museumofdesign.org

New Museum
235 Bowery, New York
NY 10002
newmuseum.org

Smithsonian Institution
PO Box 37012
SI Building, Room 153, MRC 010
Washington D.C. 20013-7012
si.edu

UC Davis Design Museum and Design Collection
145 Walker Hall, University of California
One Shields Avenue, Davis
CA 95616
designmuseum.ucdavis.edu/index.html

INDEX

Numbers in *italics* refer to captions

3D printing/prototypes 9, 110–11, *113*, 151
4c (Numnuts designers) 114–15
access (glassware) 66, *66*

A
AI 13, 92, 132
    generative AI 53
    language models for research 55–6
Alessi 144
AMO/Rem Koolhaas Roadmap 2050 Project 168–9, *169*
Apple 56, 102
architecture awards, international 168

B
Barber, Edward *101*
Barthes, Roland 56
Be your customer/client 77
Berghaus
    backpack case study 134–5, *134*, *135*
    company objective 134
Bieling, Tom *108*
bodystorming 110–11, *111*, 118
boundary users 144
Brain Health Monitoring Platform 152–3, *152*, *153*
brainstorming 60
brands
    and DNA analysis 78–9, *79*
    brandscape 80, 84
    brandscaping 79
    promoting with sports stars 147
Broken Nature (MoMA) 170
Burks, Stephen *99*
Buzan, Tony 60

C
CAD (computer-aided design) *64*, 99, 164
    drawings *38*
    images *128*
    models *101*, *107*, *112*
    prototypes *101*
    sketch modelling 99
capability simulator *see* empathy tools
Caravana (mobile coffee unit) 64–5, *64*
case studies
    4c Design's Numnuts 114–15
    AMO/Rem Koolhaas Roadmap 2050 Project 168–9, *169*
    Berghaus Freeflow: redefining backpack comfort 134–5, *134*, *135*
    delaO Design Studio's Caravana (mobile coffee unit for social impact) 64–5, *64*
    Enhancing Paper Device Usability & Design Through Task Analysis 136–7
    Giovanni Innella & Gionata Gatto's GeoMerce 170–1
    Jinil Park's Drawing Series 40–1, *40*
    Kate Strudwick's For.Form: redefining forensic evidence handling 90–1, *91*
    Lotus Theory 1 150–1, *151*
    Mischer'traxler studio's *access* (glassware) 66–7, *66*
    PA Consulting & Cumulus's Brain Health Monitoring Platform 152–3, *152*, *153*
    Parsons & Charlesworth's Catalog for the Post-Human 92–3
    Paula Zuccotti's Future Archaeology of a Global Lockdown 42–3
    Pili Wu's Plastic Ceramics 116–17, *117*
chairs
    CAD images for 128
    Cobi (PearsonLloyd) *38*
    hospital (PearsonLloyd) *126*
    Pavilion (Barber and Osgerby) *101*
    safety testing 130
    skeletal structure for 120
    testing with weights 129
Chapman, Colin 150
character profile *see* personas
ChatGPT (AI language model) 55
checklists, *see* product – checklist, PDS
circular design 12, 34
    testing 132–33, *133*
Claude (AI language model) 55
clay models *see* models
clickstream analysis 126, 128
CMF design 106
Cobi Chair *see* PearsonLloyd
Coca-Cola 56
collaboration 88, 166
collaborative (co-)design 17, 90, 166
collaging *see* product
colour trends, forecasting 82
communicating and presentation skills 159, 167
'Cone of Possibility' 33, *33*
consumer trends 86
Conterie, Punta 66
Covid-19 42
    & lockdown 42–3
critical design 162
crowdsourcing 146–7, *146*
crowdvoting 146
CSIRO 114
cultural probes (& kits) 49–50, *49*
    Probe Tools *50*
culture
    cultural comparisons & influences on design 56
    cultural observation 53
    groups, & worldwide variations 54
Cumulus (& PA Consulting) 152–3
customers/clients, understanding your 77, *77*

D
Dall'Olio, Andrea *82*
data, using and evaluating 88–9
    visualization graphs *89*
data analytics 17, 88
day in the life, A 27–8, *28*
    tutorial 46–7
Dean, Lionel T. *113*, *165*
DeepSeek (AI language model) 55
Defankle Innovation Cards 9
delaO Design Studio 64–5, *64*
Design HOPES *133*
Design Partners *112*
design proposals, evaluating and selecting 141
Design Research for Change 53
design scenarios, speculations & preferable futures 32–3, *32*
digital ethnography 22
disabilities/impaired hearing, designs for 108–9, *109*
discursive design 123
drawing from memory 81
Drawing Series 40, *40*
Droog Design 144
drop testing 130
Dunne, Anthony *49*
Dyson DC01 (vacuum cleaner) 102

E
Edison, Thomas 144
EEG brain monitoring 152
empathy tools ('capability simulators') 108–9
end-users 12, *55*, 56, 58, 82, 102, 105, 114, 123, 139, 142, 159
engaging the public, from purpose to documentation 166
Entropia Light *113*, 165
ENV motorcycle 106
environmental issues 12, 129, 132, 133, 141, 170, 177
ergonomics 99, 100, 106, *107*, 111, 114, *126*, 150, 151
ethics 17–19, 171
ethnography 21–2
　digital 22
　rapid 22
　tutorial 44–5
　visual 29
experience prototyping 105, *124*
　tutorial 118–19
external decision making 143–4, *143*
extreme users 87–8, 144

F
Facebook 146–7
Fiorelli 147
FIRA (Furniture Industry Research Assoc.) 132
FMEA (Failure Modes and Effects Analysis) 120
focus groups 73, 74–5, *74*, *78*, 85, 104, 143
forecasting *see* future forecasting
For.Form 90–1, *91*
Forms of Drinking 66
furniture 30, 40–1, 58, *99*, 100
future
　forecasting 30–1, *30*, *31*, 32, 170, 171, 177
　preferable futures, design scenarios & speculations 32–3, *32*
Future Archaeology 42–3
futurologists 30

G
Gatto, Gionata & Giovanni Innella's GeoMerce 170–1, *170*
Gaver, Bill *49*
GeoMerce 170–1, *170*

H
Handmade Furniture tables 99
Harper, Tom *80*
hearing, impaired *see* disabilities
Hemmert, Fabien *124*
Heskin , Leonie *62*
Hippel, Eric von 87
Home Interior Trend 82
Howe, Jeff 146
Hurtado, Nuria Mora *100*, 173

I
idea generation 110
ideal point 85,
IDEO 109
Ikea 53
image boards *see* mood boards
IMB 102
inclusive design 86, 108–9, 143
Innella, Giovanni & Gionata Gatto's GeoMerce 170–1, *170*
internet searches and generative AI 53–5
interviews 76, *76*
　extreme user 87–8
　'lead user' 87–8
　tutorial 96–7
'Intimate Mobiles' *124*
intuition 144
iterative design 104

J
Jouin, Patrick 111
Jung, Carl 144

K
Kane, Maire *137*
Keith, Lauren *175*
Koolhaas, Rem/AMO Roadmap 2050 Project 168–9, *169*
Kundalini *113*

L
Lamb, Max 144
literature reviews 52, *52*, *see also* reports
　tutorial 68–9
Logitech headsets *112*
logos *78*, *79*, 80, 132
Lotus 150–1, *151*
Lovegrove, Ross *76*
lowriders 53

M
MacCurrach, Rosie *82*
Management Science (journal) 87
maps & mapping, *See also* tutorials
　concept mapping 61
　journey mapping 88
　mind mapping 59, 60–1, *71*
　　tutorial 70–1
　perceptual mapping 84–5, *84*
market positioning 84
market research 80–1
Mathieson, Eilidh *163*
matrix evaluation 148–9, *149*
　tutorial 154–5
Merendi, Marco *76*
metric analysis 127
Microsoft 102
Miro 70
Mischer'traxler studio 66, *66*
mobile phones 56, 62, *120*, *124*, *see also* smartphone
　mock-up for *100*
models/modelling 99, 164, *see also* CAD; presentations; prototypes
　appearance models 106–8, *107*
　cardboard/clay/polystyrene models *99*, 100, *103*, 106, *107*, 150
　proof of concept models 101
　scale models 99–100, *99*,
　sketch modelling 99–100, *99*, *150*, 164
mood boards 82–4, 86, 146
Moredun Research Institute 114
Morning Ritual™ 92
MycoPops™ 92, *93*

N
name swapping 80
Nike 56
NootDial™ 92, *93*
Novak, Joseph D. 61
Numnuts 114–15

O
Osgerby, Jay *101*

P
PA Consulting (& Cumulus) 152–3
Pacenti, Elena *49*
paper prototyping 102, *103*, *see also* prototypes/prototyping
PAPR (powered air-purifying respirator) 136–7
Park, Jinil 40, *40*

Parsons & Charlesworth 92–3
participatory design 143, 166, see also engaging the public; stakeholder consultation
Pavilion Chair *100*
PearsonLloyd
    Cobi Chair *38*
    commode *127*
    hospital chair *126*
pen portraits *see* personas
personal belongings 29–30
personas ('character profiles'/'pen portraits') 85–6
phase zero 11
photos *21, 22, 22, 24, 24*
plastic ceramics 116–17, *117*
presentations, *see also* reports
    guidelines 164
    preparing 159–61, *159, 160*
    top tips for 160
    tutorial 172–3
    visuals and models 164, *165*
product
    autopsy 34, *35*
    camouflage 80
    champion 147–8, *147*
    checklists 142–4, *see also* PDS (below)
    collage 86–7, *86*
    competitor analysis 49, 51, *51*
    design, key points 176–7
    life cycle of *16*
    PDS (product design specification) 11, 141, *141*, 142, *142*, 14–9, 154, 155
    tutorial 156–7
    usability testing 102, 126–77, *136*
        metrics analysis 127
        RITE 127
proof of concept 101
prototypes/prototyping 110, *121*, *see also* individual entries
    quick-and-dirty, tutorial 120–1,
    versus paper prototyping 102
public engagement 166
Pugh, Stuart 148, *see also* matrix evaluation

Q

questionnaires & surveys 73–4, *73*, *see also* interviews; public engagement; reports
    tutorial 94–5
quick-and-dirty prototypes 104, *104, 105*, *see also* tutorials

R

rapid prototyping 110–11, *111*
reflection, importance of 19
regenerative design/practice 33, 99
RemWake *92*
Renault 5, electric 29
reports, how to write 162–3, *163*
    tutorial 174–5
research, *see also* literature reviews; presentations; questionnaires & surveys; reports
    analysing 16–17
    defining 9
    and design 10, *16*
    and digital technologies *11*
    ethics 12, 17–19, 86
    iterative design process 11–12, *13*
    market and retail 80–1
    people behaviour & products 49
    practice-based 10, 11
    primary 10, 10–11, 28, 52, 134
    product design methods 12–13, *22*
    and reflection 19
    secondary 10–11
    scientific versus design 9–10
    summary of methods 14–15
RITE (rapid iterative testing and evaluation) 126–8, *126*
role playing 49, *55*, 56–8, *74, 136*

S

safety, *see also* testing
    safety standard marks *130*
sampling *60*, 61–2, 88
scenarios 30–1, 32–3, *32*, 57, 77, 88, 136, 171
    testing 123
Schön, Donald 19
'Sensory Threads' *118, 119*
shadowing 26, *27*, 28, 47, 64
sketch models *see* models/modelling
sketching 37–8, 164
    concept *36, 37, 38*, 99, *99, 103*
    schematic 38
    thematic 37
Skins, Sonia 43
SMART principles 58
smartphone *51*, 56
social media 31, 146, *147*
speculations, design scenarios & preferable futures 32–3, *32*
speculative design 32–3, 170

stakeholder consultation, engaging with 30, 37, 57, 82, 88, 90, 105, 114, 120, 123, 124, 136–7, 139, 143, 147, 159, 171, 172, *see also* external decision making
stereolithography 111
storytelling *29*
Strudwick, Kate 90–1
subcultures 56
surveys *see* questionnaires & surveys

T

task analysis 62–3
    Earee *62*
    octoplasty surgery *62*
testing, *see also* mock-ups; models/modelling
    A/B 28, 126, 128
    Berghaus *see* case studies
    circular design testing 132–3
    drop testing *130*
    material testing 116, 129–30, *129*
    PearsonLloyd chair, commode *124, 125*
    safety testing 130–2, *130*
    remote testing 126, *127*
    RITE (rapid iterative testing and evaluation) 126–8, *126*
    scenario testing 123
    test rigs *100, 101*, 114, *115*, 123, 129–30
    type testing 132
    user trials 124, *124*
Testori, Elisa 66
theatre caps *133*
'think aloud protocols' 75, *see also* user narration
touchpoint analysis 81, *81*
trend forecasting 30
trend spotting 31, *see also* forecasting and online forecasting *31*
Triennale Design Museum 37
'Try it yourself' 58, *58*
tutorials
    How to conduct a day-in-the-life study 46–7, *46*
    How to conduct an ethnographic study 44–5, *45*
    How to conduct experience prototyping 118–19
    How to conduct great interviews 96–7

How to conduct a matrix evaluation 154–5, *155*
　How to create a great mind map 70–1
　How to create a great questionnaire 94–5, *95*
　How to create a great research presentation 172–3, *173*
　How to create a great research report 174–5, *175*
　How to do quick-and-dirty prototyping 120–1, *120*
　How to run a great user trial 138–9, *139*
　How to write a checklist (PDS) 156–7
　How to write a literature review 68–9, *69*

U
unfocus groups 75, *see also focus groups*
urban tribes 56
Urushi Stool *144*
Use Your Head 60
user narration 75
user trials 123, 124, *124*
　tutorial 138–9

V
video/video diaries 22, *23*, 24, *24*
Voros, Joseph *33*

W
Wanders, Marcel 111
Weir, Johnny *96*
Which? (magazine) 51
Whyte, Gregor *34, 36, 38, 104*
Wired (magazine)146
Wordle™ 59
writing reviews, tutorial 68–9
Wu, Pili 116–17, *117*

X
X (formerly Twitter) 147

Z
Zuccotti, Paula *29*, 42–3, *43*
　Every Thing We Touch 42

PICTURE CREDITS

The authors and publisher would like to thank the following institutions and individuals who provided images for use in this book. In all cases, every effort has been made to credit the copyright holders, but should there be any omissions or errors the publisher would be pleased to insert the appropriate acknowledgement in subsequent editions of this book.

a = above, b = below, c = centre, l = left, r = right

cover Drawing Series © JINIL PARK
8, 9 Dr Abi Hird, Founder & Director, Defankle Innovation Limited
11 Moment Makers Group/istockphoto
13a, 13b, 14, 15, 16 Hazar Taissier Marji
20 Courtesy of Paula Zuccotti from her project 'Every Thing We Touch'
21 Courtesy of James Shutt
22 Courtesy of IDEO
23 Courtesy of Ana Alves and Rui Alves, University of Madrid
24, 25 Courtesy of Northumbria School of Design
27 Courtesy of James Shutt
28 © James Leynse/Corbis
29 Courtesy of Paula Zuccotti from her project 'Every Thing We Touch'
30 © OMA
31 trendsenses.com
32 Courtesy of Isaac Teece
33 Hazar Taissier Marji
34 © Gregor Whyte 2025. All rights reserved
35 Courtesy of Aesir, photographer Jonathan de Villiers
36 © Gregor Whyte 2025. All rights reserved
37 Courtesy of La Triennale Design Museum, Martí Guixé sketch
38 © Gregor Whyte 2025. All rights reserved.
39 Courtesy of PearsonLloyd
41 Drawing Series © JINIL PARK
43 Courtesy of Paula Zuccotti from her project 'Every Thing We Touch'
45 © Josef Mohyla/iStockphoto
46 © Pedro Castellano/iStockphoto
48, 49, 50 © Interaction Research Studio

51 Bet Noire/istockphoto
52 © urbancow/iStockphoto
53 Paul Rodgers
54 © Frank Delm/Getty Images
55 Courtesy of University of Northumbria School of Design
58 © George Peters/iStockphoto
59a, 59b Hazar Taissier Marji
60 © TommL/iStockphoto
62, 63 Leonie Heskin Professor of Simulation-based Education and Director of ASSERT Simulation Centre; NCAD MSc Medical Device Design students Alexa Broen, Aoife Gallagher and Lochlann O'Regan; and NCAD design staff Enda O'Dowd and Derek Vallence
65 Images by delaO design studio
67 mischer'traxler studio, 2022, for Punta Conterie, produced by Vetreria Simone Cenedese and Eugenio Panizzi, image Francesco Allegretto
69, 71 Hazar Taissier Marji
72 Linnemann Design and Kenoteq Ltd
74 Courtesy of University of Northumbria School of Design
75 © Erik Bohemia courtesy of University of Northumbria School of Design
76l Courtesy of Rory Hyde
76r Courtesy of Vola Nito Raspel
77 © Joos Mind/Getty Images
78 Fieldwork/istockphoto
79 Courtesy of Isaac Teece
80 Courtesy of Tom Harper, Edinburgh College of Art
81 Hazar Taissier Marji
82 Courtesy of Rosie MacCurrach
83a Photograph by Mischa Haller www.mischaphoto.com
83c, 83b Andrea Dall'Olios's Home Interior Trend book, S/S 2010
84 Hazar Taissier Marji
86 Photograph by Mischa Haller www.mischaphoto.com
89 gapminder.org
91 Kate Strudwick
92, 93 Photo Parsons & Charlesworth
96 Courtesy of Jonny Weir
98 Nuria Mora Hurtado, Product Design & Innovation graduate
99 Part Tables, Stephen Burks
100 Nuria Mora Hurtado, Product Design & Innovation graduate

101al, 101ar Edward Barber and Jay Osgerby's De La Warr Pavilion Chair
101c, 101b Courtesy of University of Northumbria School of Design
103 Courtesy of Scholten & Baijings
104 © Gregor Whyte 2025. All rights reserved
105 Courtesy of University of Northumbria School of Design
106, 107 Courtesy of Seymourpowell
108, 109 Tom Bieling, Design Research Lab
110 Deepak Sethi/istockphoto
111 Marina Skoropadskaya/istockphoto
112 Courtesy of Design Partners
113 © Lionel T. Dean
115 4c Design Ltd (Glasgow)
116, 117 Courtesy of OXO Good Grips
118, 119 Courtesy of Proboscis
120 Courtesy of Deutsche Telekom Laboratories
121 Courtesy of Wataru Watanabe
122 Berghaus Ltd
124 © BMW AG
125 Courtesy of Design Research Lab & Deutsche Telekom Laboratories
126, 127, 128, 129 Courtesy of PearsonLloyd
131a © Maurice Volmeyer/Shutterstock
131b © Vladislav Gajic/Shutterstock
133 Hazar Taissier Marji
134, 135 Berghaus Ltd
136, 137 Máire Kane – Medical Device Designer, Enda O'Dowd – Joint Course Coordinator MSc Medical Device Design, Derek Vallence – Joint Course Coordinator MSc Medical Device Design
139 4c Design Ltd (Glasgow)
140 PA Consulting design and innovation team
141, 142 Hazar Taissier Marji
143 Courtesy of Will Mitchell
145 Courtesy of Max Lamb
146 titoslack/iphotostock
147 SOPA Images Limited/Alamy Stock Photo
149 Hazar Taissier Marji
150, 151 Courtesy Lotus Cars
152, 153 PA Consulting design and innovation team
155, 156 Hazar Taissier Marji

158 Photo of the GeoMerce installation as exhibited during the Design Week in Milan, 2015. Photo by Matteo Cremonini. Courtesy Gionata Gatto and Giovanni Innella.
159 Courtesy of Propeller Design Team and Kapsel Multimedia AB
161 Courtesy of Seren
163 Eilidh Mathieson (referenced third-party materials are cited within the project's original pages under academic fair use)
165 Entropia Light by Lionel T. Dean
169 © OMA
170, 171 Photos of the GeoMerce installation as exhibited during the Design Week in Milan, 2015. Photos by Matteo Cremonini. Courtesy Gionata Gatto and Giovanni Innella.
173 Nuria Mora Hurtado, Product Design & Innovation graduate
175 Lauren Keith, MEng Product Design Engineering

CASE STUDY CREDITS

Drawing Series designed by Jinil Park, a South Korean designer and artist.

Future Archaeology by Paula Zuccotti, a London-based Argentine industrial designer, ethnographer, trends forecaster, creative strategist and visual artist.

Caravana by delaO design studio, based in Mexico. Client: Cirklo, Citibanamex Compromiso Social, Union Mahomut. Brand Identity: Cirklo. Creative Direction: José de la O. Product Design: Montserrat Pazos, Rodrigo Piña & José de la O. Research: Montserrat Pazos & José de la O. Images and Video: Montserrat Pazos, Sander Verbeek, Rodrigo Piña & José de la O.

Access, designed by mischer'traxler studio based in Vienna, Austria. Produced by glass craftsmen Vetreria Simone Cenedese and Eugenio Panizzi, the project was commissioned as a limited edition by Punta Conterie for the exhibition *Forms of Drinking*, curated by Elisa Testori.

For.Form by Kate Strudwick, a multidisciplinary designer focused on building bridges between technology and human problems using design research, user experience and visual design. Kate is currently an interaction designer at Google Playspace Labs in San Francisco, USA.

Catalog for the Post-Human by Parsons & Charlesworth, a design studio founded in 2014 by Chicago-based British husband-and-wife team Tim Parsons and Jessica Charlesworth.

Numnuts designed by 4c, a design studio based in Glasgow, in partnership with the Moredun Research Institute (MRI) in Scotland, Meat and Livestock Australia (MLA), CSIRO and Australian Wool Innovation (AWI) in Australia.

Plastic Ceramics designed by Pili Wu for Lyngby Porcelain, in partnership with the Han Gallery, Taipei.

Berghaus Freeflow designed by the Berghaus product development team based in the North East of England, UK.

Powered Air-Purifying Respirator designed by Máire Kane, under the supervision of Enda O'Dowd and Derek Vallence, NCAD, Dublin.

Lotus Theory 1 designed by the Lotus Design Team under the leadership of Ben Payne, Vice President of Design, Lotus Group.

Cumulus EEG Headset designed by PA Consulting (Dublin Office) and Cumulus Neuroscience. The team included: Terence Kealy (Industrial Designer), Sara Urasini (Industrial Wearables Designer), Eugene Canavan (Human Factors Expert), Bryan Murphy (Mechanical Engineer), Glyn Griffiths (Electronics Engineer), Eoin McNally (Industrial Designer), Ian McCullough (Cumulus Neuroscience Team), Giedrė Kaktytė (Cumulus Neuroscience Team), Mark Armstrong (Cumulus Neuroscience Team) and Brian Murphy (Cumulus Neuroscience Team).

RoadMap 2050 by AMO, the research, branding and publication studio of the architectural practice Office for Metropolitan Architecture (OMA), which was founded in London in 1975 by Rem Koolhaas and others, and relocated to Rotterdam in 1978.

GeoMerce concept and design by Gionata Gatto and Giovanni Innella. Electronics and mechanics by Eelke Feenstra. Video and animations by Max Italiaander. Photos by Matteo Cremonini. Partners: WUR (Plant Sciences Group Wageningen, Life Sciences Department of the University of Parma, LINV (International Laboratory of Plant Neurobiology); C-CIT. Funded by Stimuleringsfund Creative Industry; Dutch Embassy in Milan (Italy).

ACKNOWLEDGEMENTS

We would like to express our gratitude to our colleagues and students at the National College of Art and Design, Ireland, and the University of Strathclyde, Scotland.

We would also like to thank all of those designers who contributed their wonderful images of work and in particular to those who also gave their time to discuss and provide constructive thoughts and feedback as the book developed. Alex Milton would especially like to thank Suzanne and Matilda for their support and understanding. Paul Rodgers would especially like to thank Alison, Charlie and Max for their continual support.

Finally, we would like to thank Kara Hattersley-Smith, Liz Faber and the team at Laurence King for helping us to complete the second edition of this book.